# SpringerBriefs in Applied Sciences and Technology

For further volumes:
http://www.springer.com/series/8884

Jinghai Li · Wenlai Huang

# Towards Mesoscience

## The Principle of Compromise in Competition

 Springer

Jinghai Li
Wenlai Huang
Institute of Process Engineering
Chinese Academy of Sciences
Beijing
People's Republic of China

ISSN 2191-530X          ISSN 2191-5318   (electronic)
ISBN 978-3-642-41789-4     ISBN 978-3-642-41790-0   (eBook)
DOI 10.1007/978-3-642-41790-0
Springer Heidelberg New York Dordrecht London

Library of Congress Control Number: 2013953642

Printed on acid-free paper

Springer is part of Springer Science+Business Media (www.springer.com)

*A brief for*

*"From Multiscale Modeling to Meso-Science
— A Chemical Engineering Perspective"*

*Springer, 2013*

# Preface

Recently, Springer published a monograph entitled: "From Multiscale Modeling to Meso-Science—*A Chemical Engineering Perspective*," which summarizes three decades of research on mesoscale modeling, mainly in chemical engineering. At the end of the book, the new concept of mesoscience was proposed based on the observed generality of different systems, but was not discussed in detail.

From the feedback of readers, we recognized that the original book helps readers to understand mesoscale modeling and simulation, but does not reveal the whole picture of mesoscience because it only provides a very simple description. In addition, we have further clarified the concept of mesoscience since the publication of the book. It is therefore necessary to publish a brief to extract the most relevant content regarding mesoscience to provide a more comprehensive description of this concept.

Supported by Springer, this brief introduces the evolution of the principle of compromise in competition. We outline how the multiscale modeling of particle clustering in gas–solid systems was developed, verified, extended, and finally generalized into the energy-minimization multiscale (EMMS) principle of compromise in competition for all mesoscale phenomena. In addition, more information on the concept of mesoscience are supplemented with our increased understanding based on new evidence acquired, particularly, the understanding of stability conditions in protein foldings, turbulence, and materials preparations. We hope this brief will allow readers to readily understand the concept that we term mesoscience. If mesoscience can be established as a general interdisciplinary science, it may allow us to overcome many current research challenges. This is the driving force that encouraged us to write first the book and now this brief.

We express our sincere thanks to the authors of the original book and all colleagues in the EMMS group. Special thanks are extended to Springer for the encouragement to write this brief.

# Contents

# Symbols

| | |
|---|---|
| $C_D$ | Drag coefficient, dimensionless |
| $d_{cl}$ | Cluster diameter, m |
| $d_e$ | Diameter of eddies, m |
| $d_i$ | Diameter of entity $i$ (large/small bubble), m |
| $E_j$ | Objective function with respect to dominant mechanism $j$ |
| $f$ | Volume fraction of dense phase, dimensionless |
| $f_e$ | Volume fraction of eddies, dimensionless |
| $f_i$ | Volume fraction of entity $i$ (large/small bubble), dimensionless |
| $F$ | External force field, m/s$^2$ |
| $F_i$ | Constraint condition $i$ |
| $H_A$ | Potential A in unit mass, J/kg |
| $H_B$ | Potential B in unit mass, J/kg |
| $L$ | Distance, m |
| $m$ | Mass, kg |
| $N_{surf}$ | Rate of energy dissipation caused by bubble breakage and coalescence per unit mass, W/kg |
| $N_{st}$ | Rate of energy consumption for transporting and suspending particles per unit mass, W/kg |
| $N_{turb}$ | Rate of energy dissipation in turbulent liquid phase per unit mass, W/kg |
| $r$ | Radial coordinate, m |
| $R$ | Radius, m |
| $R_m$ | Mass ratio, dimensionless |
| $R_v$ | Ratio of random velocity to directed velocity, dimensionless |
| $U$ | Superficial velocity, m/s |
| $U_c$ | Superficial velocity of gas in the dense phase, m/s |
| $U_f$ | Superficial velocity of gas in the dilute phase, m/s |
| $U_g$ | Superficial velocity of gas, m/s |
| $U_{g,i}$ | Superficial velocity of gas in entity $i$ (large/small bubble), m/s |

| | |
|---|---|
| $U_l$ | Superficial velocity of liquid, m/s |
| $U_p$ | Superficial velocity of solid, m/s |
| $U_{pc}$ | Superficial velocity of solid in the dense phase, m/s |
| $U_{pf}$ | Superficial velocity of solid in the dilute phase, m/s |
| $v$ | Random velocity, m/s |
| $V$ | Directed velocity, m/s |
| $W_{st}$ | Rate of energy consumption for transporting and suspending particles per unit volume, W/m$^3$ |
| $W_{te}$ | Rate of energy dissipation for inertia per unit volume, W/m$^3$ |
| $W_v$ | Rate of energy dissipation for viscosity per unit volume, W/m$^3$ |
| $\mathbf{x}$ | State vector |
| $x_i$ | Component $i$ of state vector |
| $\alpha$ | Angle, ° |
| $\varepsilon$ | Local average voidage, dimensionless |
| $\varepsilon_c$ | Voidage in dense phase, dimensionless |
| $\varepsilon_f$ | Voidage in dilute phase, dimensionless |
| $\mu_e$ | Viscosity of fluid in eddies, Pa·s |
| $\mu_l$ | Viscosity of liquid, Pa·s |
| $\mu_{lam}$ | Viscosity of fluid in laminar flow, Pa·s |
| $\rho_e$ | Density of fluid in eddies, kg/m$^3$ |
| $\rho_g$ | Density of gas, kg/m$^3$ |
| $\rho_l$ | Density of liquid, kg/m$^3$ |
| $\rho_{lam}$ | Density of fluid in laminar flow, kg/m$^3$ |
| $\rho_p$ | Density of particles, kg/m$^3$ |
| $\sigma$ | Surface tension, N/m |

# Chapter 1
# Background: Correlating Microscale and Macroscale

## 1.1 A Common Challenge

Since the second half of the twentieth century, our knowledge of microscale information and macroscale behaviors of complex phenomena has expanded, and researchers now anticipate complete solutions to various challenging problems. However, it remains difficult to correlate system behaviors with microscale information, such as organism characteristics via genomics, protein characteristics via proteomics, turbulence through molecular interactions, and superconductivity considering a single electron. That is, reductionism does not currently meet with holism. What is missing in between? Bridging the gap between these two well-understood areas is a challenge in contemporary science and technology. We believe that the gaps are caused by an overlooked common governing principle that is valid at various mesoscales. In addition, the term mesoscales itself is not widely understood. This provided us with the driving force to write this brief, which presents a preliminary but relatively complete introduction to the concept of mesoscience.

In practical engineering, even though we understand all of the elemental details involved in a specific system we are still unable to optimize the performance of processes with respect to energy, resources, and the environment. The collective behaviors of elements are critical to the overall properties of a system but are very complicated to understand, describe, and predict.

The situation is the same in chemical process engineering. For example, we are unable to design materials from the properties

J. Li and W. Huang, *Towards Mesoscience*, SpringerBriefs in Applied
Sciences and Technology, DOI: 10.1007/978-3-642-41790-0_1,
© The Author(s) 2014

of molecules, and reactor properties cannot be predicted from our understanding of single particle behavior. Therefore, scaling up chemical processes from laboratory to industrial scale remains a challenge, implying that our understanding of the mesoscales at material and reactor levels between element and system scales is limited.

As a result, researchers in various fields have looked for possible theories and methodologies to correlate microscale and macroscale; that is, to understand collective behaviors of elements, as explored in complexity science [1]. However, progress in this direction has not been satisfactory. Different groups have failed to recognize the similarity between systems, and have explored the disparity between scales from different angles, leading to a variety of concepts including complexity, nonlinearity, self-assembly, self-organization, emergence, dissipative structure, chaos, intermittency, and order and disorder. Although interest in reaching the common goal of correlating microscale and macroscale is increasing, confusion still exists over the best approach to explore this problem. It is difficult to say why the exploration of complex systems progressed slowly, but the ignorance of the existence of the governing rule at mesoscales might be one of the reasons.

## 1.2  Questions and Opportunities for Engineers

With our current inability to bridge the gap between microscale and macroscale, it is rational to ask: what have we overlooked? What is the right angle to view this problem? Is there a governing principle that will allow us to correlate microscale information and system behaviors? Is there a unifying relationship between the established concepts? These questions are applicable to all fields because of the generality of this problem. However, they can be answered only by studying individual systems to extract common factors because of the diversity and complexity of the systems, which gives engineers the opportunity to contribute to a general interdisciplinary science.

This is different from traditional sciences, which were usually established to address a specific field, and then applied to others. For instance, the laws of transport phenomena originated from chemical engineering, but have been applied to many different fields such as energy, resource, and environment, and even biological technology.

Computational science was generated as a specific science, but has been widely used in the whole spectrum of science and technology.

Engineers traditionally solve engineering problems using available knowledge from basic sciences. For example, chemical engineers use knowledge of chemistry, physics, biology, and mathematics to develop new chemical and biological technology, processes and equipment for energy, materials and other industries. However, because of new challenges, engineers have recognized that current knowledge is insufficient to solve many new engineering problems. They also observed that most gaps in their knowledge appear at scales between elements and systems. Therefore, we inferred that there may be a common principle governing these limitations, which has stimulated engineers to explore new principles and upgrade their knowledge base. This brief and our previous book describe this type of research performed by chemical engineers; that is, bottom-up research of different systems to extract common principles.

## 1.3  An Interest in Mesoscales

"Meso" is derived from the ancient Greek word *mésos*, meaning "middle," and is used to describe the scale between small and large; that is, intermediate phenomena. While "mesoscopy" normally means a specific range of physical size between 100 and 1,000 nm [2], where quantum mechanics meets classical mechanics, "mesoscales" discussed here is a relative concept, and refers to the ranges between the element and system scales, which is not limited to a specific range of physical size but can be any subrange of physical size between elementary particles and the observable universe [3], as summarized later in Fig. 6.3. Therefore, the concept of "mesoscales" is related to the whole spectrum of science and technology. Mesoscales can be observed only when we resolve the fine details of systems, otherwise, two or even more mesoscales could be blurred, leading to confusion on the governing rules. Mesoscale phenomena are always related to the collective behavior of elements showing heterogeneity or dynamic changes.

We became involved in this direction of research unintentionally. In the 1980s, we studied particle clustering—which occurs at the mesoscale between single particle and system—in gas–solid fluidization because of its influence on mass transfer and reaction processes. This

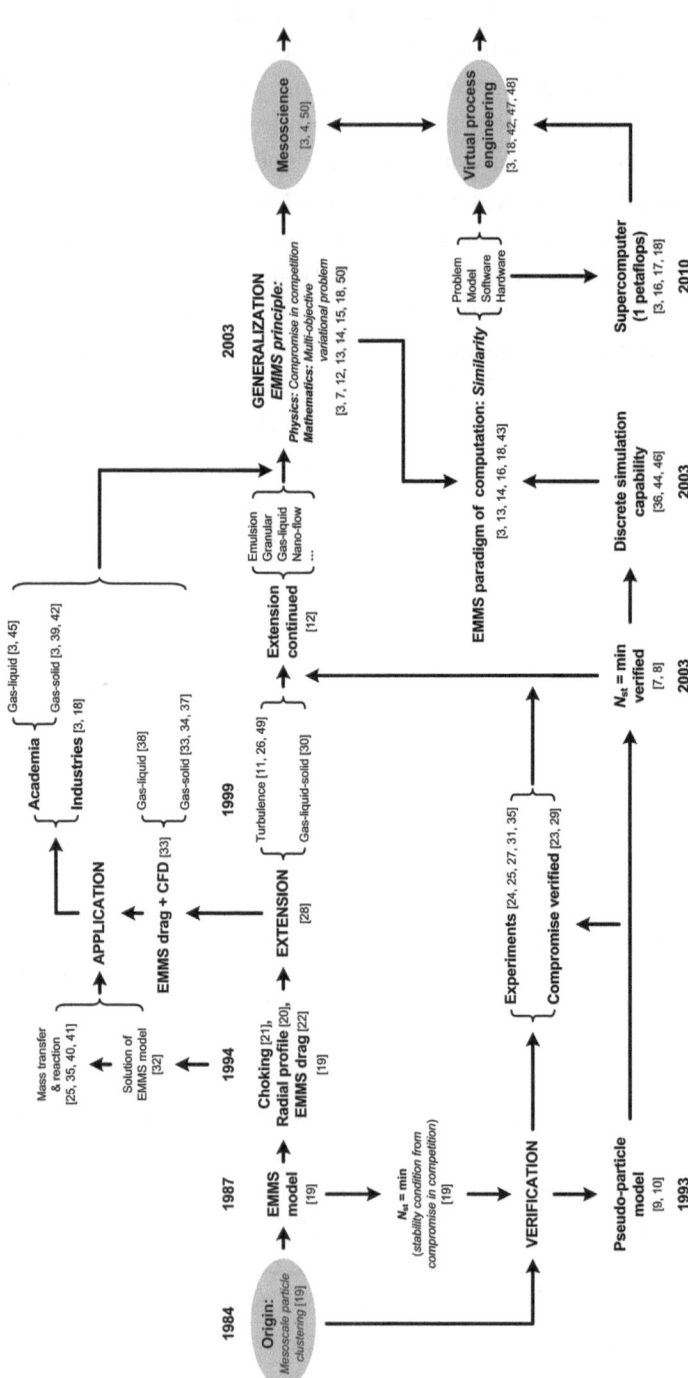

**Fig. 1.1** Flow chart *highlighting* three decades of multiscale research at IPE, CAS. Modified from [3], Copyright 2013, with permission from Springer

research led us to develop the EMMS (energy-minimization multiscale) model. The EMMS model is governed by the principle of compromise in competition between different dominant mechanisms, and is later formulated as a multiobjective variational (MOV) problem. Its extension to turbulent flow in pipes and other complex systems stimulated our interest to explore the universality of the EMMS principle. Gradually, we recognized that it may be possible to identify a common principle and establish a general formulation to correlate microscale and macroscale; that is, the potential for an interdisciplinary science: mesoscience [3, 4].

Here we show how we came to predict the concept of mesoscience and our current understanding. We hope this will trigger interest in mesoscience from researchers in different fields.

## 1.4 Footprint

Figure 1.1 gives a summary of the three decades of research on mesoscale modeling of various complex systems at the Institute of Process Engineering, Chinese Academy of Sciences (IPE, CAS) from the specific example of particle clustering to the proposition of the general concept of mesoscience. Starting with the study of particle clustering, the EMMS model [5, 6] was established to describe gas–solid fluidization systems by analyzing the compromise in competition between dominant mechanisms. The EMMS model was then verified [7, 8] using the pseudo-particle model approach [9, 10], extended to other complex systems [11, 12], and applied in different industries. Gradually, the EMMS principle, which is simply *"compromise in competition between dominant mechanisms,"* was established. This principle suggests the possible universality of all problems at mesoscales between elemental particles and the observable universe [13, 14], and can be formulated as MOV problems [15]. The common physical principle and universal mathematical formulation of the EMMS principle led us to propose the concept of mesoscience [3, 4]. In parallel, we defined the EMMS paradigm for computation, which features structural and logical consistency between the physical problem, mathematical modeling, numerical algorithm (software), and hardware [13, 14, 16]. The EMMS paradigm for computation was implemented by constructing a

supercomputer with a capacity of 1.0 petaflops [16–18]. We believe
that this paradigm will allow us to realize virtual process engineering (VPE), that is, real-time simulation. Obviously, the progress on
mesoscience will enable the computational science, and vice versa,
computational science will be a powerful tool for mesoscience.

## 1.5  Objectives

The flow chart in Fig. 1.1 shows the evolution of our study from
multiscale modeling of particle clustering in gas–solid systems to a
generalized principle of mesoscience to explore all mesoscale phenomena. In this brief, we describe the principles of mesoscience in
a systematic way. We also include our latest advances in understanding mesoscience and evidence to answer the following questions that
arose from the book: What are mesoscales? What is mesoscience?
What is the possible common principle describing all mesoscale
phenomena? Particularly, several points critical to mesoscience are
clarified with perspectives.

# Chapter 2
# The EMMS Principle: Compromise Between Competing Dominant Mechanisms

## 2.1 The Importance of Structural Heterogeneity

Dynamic heterogeneous structures that appear on the mesoscales between element and system scales make correlating microscale and macroscale difficult (Fig. 2.1). Such heterogeneous structures are currently bottlenecks that prevent us from fully understanding the behavior of complex systems in science, engineering, nature, and even society.

The universe contains heterogeneous structures that form on different levels through complex dynamic processes. According to the Standard Model of particle physics [51], elementary particles of matter are quarks and leptons (including electrons), and gauge bosons act as force carriers. Two quarks form a meson, and three quarks form a baryon (including nucleons). Therefore, mesons or baryons are structures at the lowest level known to date. Fewer than 300 nucleons form an atomic nucleus, which is the level of structure above nucleons. Atomic nuclei and electrons can form plasma, or an atomic nucleus and fewer than 120 electrons can form an atom, which are another level of structures. Atoms form molecules, macroscopic gases, or condensed matter, representing structures at even higher levels. Huge amounts of condensed matter form planets, whereas very large amounts of gases or plasma form stars. Planets form a planetary system around a star, and stars with planets form galaxies [52], which can further form groups, clusters, and even superclusters [53]. Superclusters are currently the highest level structures in the observable universe. We consider that heterogeneity (cf. Fig. 6.3) exists in most systems [7].

J. Li and W. Huang, *Towards Mesoscience*, SpringerBriefs in Applied Sciences and Technology, DOI: 10.1007/978-3-642-41790-0_2,
© The Author(s) 2014

**Fig. 2.1** Mesoscale between a single element and system

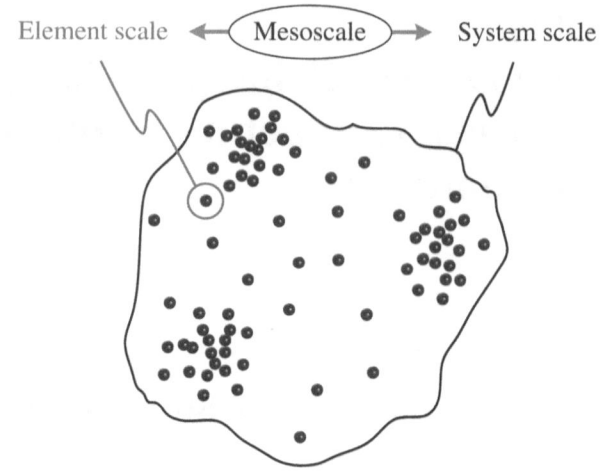

Chemistry, materials science, and condensed matter physics focus on the dependence of bulk material properties on atomic or electronic structures [54]. The structures of molecular clusters have become another focus in recent years with the development of supramolecular chemistry and nanotechnology [55]. These fields include not only the stationary structures at equilibrium state but also the dynamic processes prevailing during their formation. Geology [56] and astronomy [57] investigate the inner and outer structures of the earth and celestial objects. Meanwhile, biology [58] studies the structures of organic molecules, cells, tissues, organs, and their interdependence. Biology also probes further downwards to genes and upwards to systems.

Chemical engineering often involves dynamic structures. Dynamic structures are mostly complex, making it difficult to scale-up reactors. Without appropriate understanding of dynamic structures, quantitative design and control of processes is seldom possible. However, insufficient attention has been paid to understanding dynamic structures because no unifying theory governing all complex systems exists, and complexity science itself is still a field of "perplexity" [59]. Therefore, dynamic structures are a challenge not only for chemical engineering but also for all other fields investigating complex systems. Chemical reactors can be designed and optimized quantitatively only when various structural changes in systems can be predicted.

We initially studied particle clustering in gas–solid two-phase flows. Such particle clustering is a typical mesoscale phenomenon, featuring the interaction at or between three scales: the microscale of individual particles, mesoscale of particle clusters, and macroscale of the global system [5, 6], which was the starting point of this series of research.

The three cases of structures shown in Fig. 2.2 have the same element (particle), system volume, and number of particles. However, they have quite different mesoscale structures, which lead to different transport and reaction performance.

Both local structure and overall structure in gas–solid systems affect system properties. Figure 2.2 shows the considerable difference of transfer rate, represented here by the drag coefficient $C_D$, for different structures with the same amount of solid particles in a given volume (i.e., the same average voidage) and the same gas flow rate [15]. Local structure leads to a decrease in $C_D$ from Fig. 2.2(a) to (b), even though the average parameters for these two structures are identical. Overall structure leads to a further decrease of $C_D$ from Fig. 2.2(b) to (c), indicating that the formation of a core-annulus structure reduces the global transfer rate. The contribution of dynamic changes in a structure to mass transfer rate has been further demonstrated experimentally [25, 35]. Therefore, it is important to understand these structural changes and their influence on transport phenomena during predicting the performance of chemical reactors, which is the most challenging issue in chemical engineering.

## 2.2 Analysis of Heterogeneous Structures

Heterogeneous structures are common in all multiscale systems. They may be analyzed by one of the following three approaches:

- The first is the *discrete method*, which is based on known information at the element scale, and tries to understand complex systems at higher scales by analyzing lower-scale mechanisms. However, if the lowest-scale mechanisms are not fully understood, any deviation at this scale is magnified at higher scales. In addition, discrete

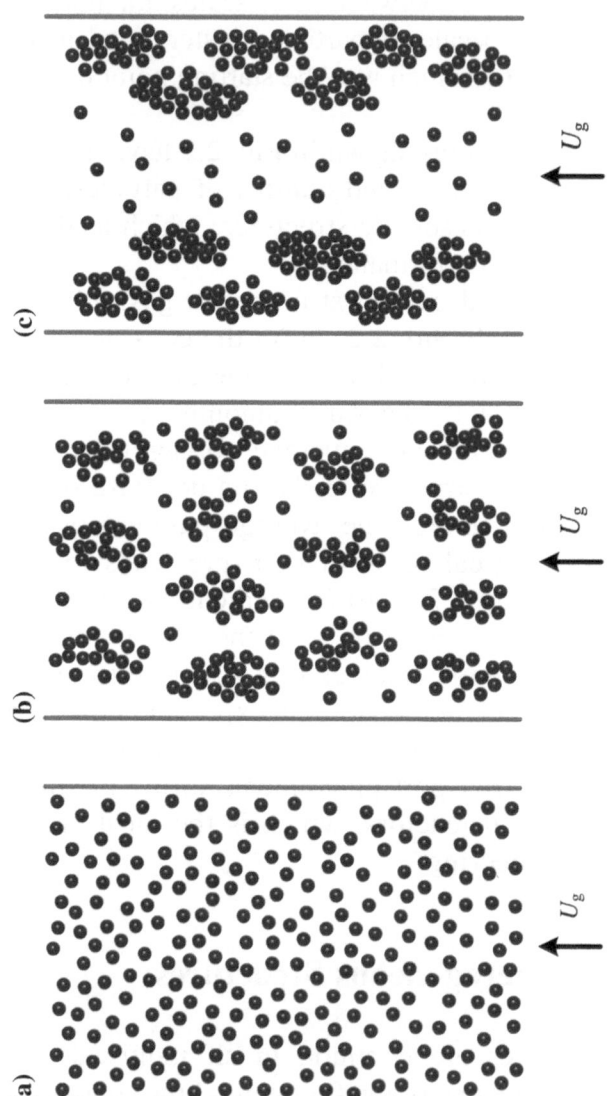

**Fig. 2.2** Critical dependence of transport behavior on mesoscale structure. **a** Without structures $C_D = 18.6$. **b** With only local structures $C_D = 5.43$. **c** With both local and global structure $C_D = 2.85$. Reprinted from [15], Copyright 2013, with permission from Elsevier

approaches have high computation cost, so it remains difficult to simulate large systems.

- The second is *averaging* all parameters over specific space or time by considering the system to be spatially or temporally uniform. This is the simplest approach and is usually practical. However, it cannot correctly describe heterogeneous structures, leading to errors.
- The third approach involves taking the *multiscale* structure into account and considering the disparity of behaviors and interactions at and/or between different scales.

Among these three approaches, the averaging one, although commonly used, is not sufficient to formulate transport phenomena in heterogeneous structures because it does not distinguish between different scales. This is shown in Fig. 2.3 for gas–solid two-phase systems consisting of a solid-rich dense phase and gas-rich dilute phase. Gas–solid interaction displays a multiscale nature involving three different mechanisms [19, 21, 28]:

- *"Particle-dominated"* in the dense phase, that is, particle movement is realized by suppressing gas movement.
- *"Gas-dominated"* in the dilute phase, which means that the gas movement is realized by suppressing particle movement.
- *"Particle-gas-compromising"* between the dilute and dense phases. Here, the movement of neither the gas nor the particles dominates, and therefore, they compromise.

If the averaging approach is used, these three different mechanisms will be blurred, and the value of $C_D$ will be distorted, as shown in Fig. 2.3. The data in Fig. 2.3 have been calculated for a particle-air system [19], where the particle diameter is 54 μm, the particle density is 930 kg/m$^3$, the solid flow rate is 50 kg/(m$^2$·s), and the gas velocity is in the range between 1.5 and 3.2 m/s.

In conclusion, most structures in chemical engineering are multiscale, which cannot be appropriately quantified by the averaging approach. The discrete approach based on microscale mechanisms are not yet practical because of the limitations of measurement technology and computer capacity. Multiscale methodology was, therefore, considered as a promising approach for describing both stationary and dynamic structures. However, a generalized methodology had not yet been developed, which stimulated this series of study at IPE, CAS in the 1980s.

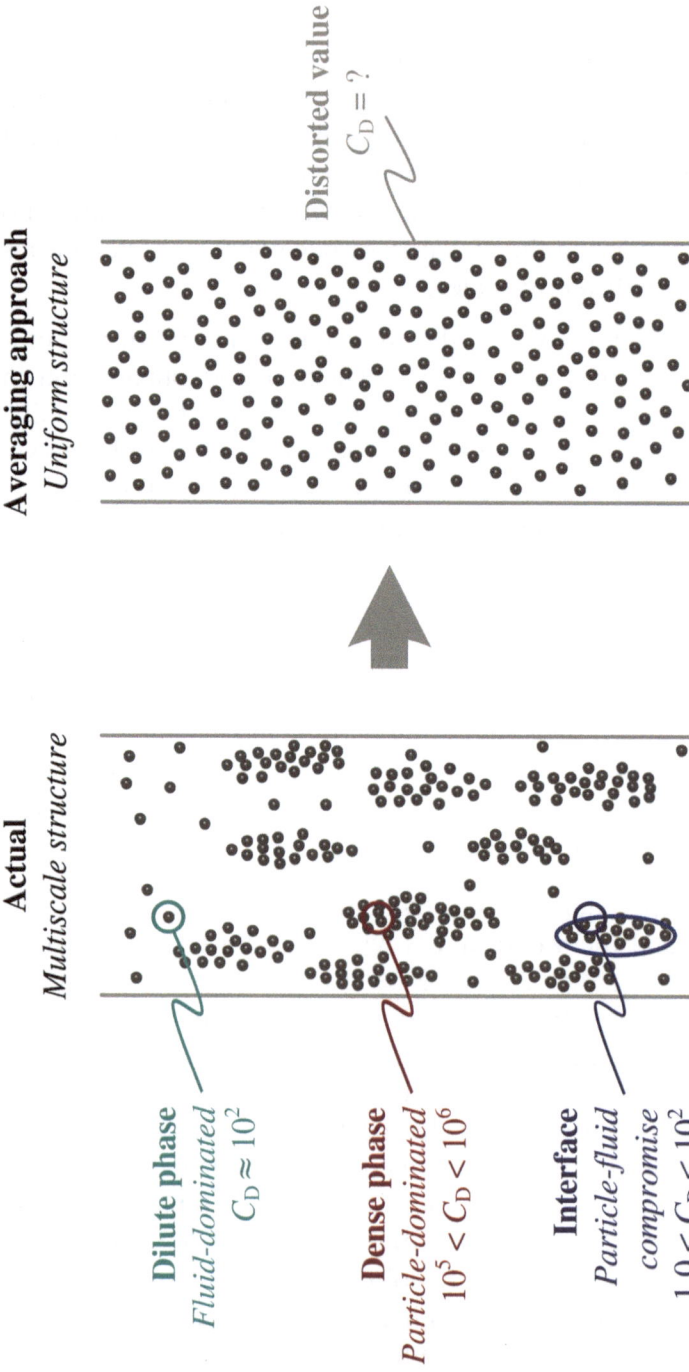

**Fig. 2.3** Insufficiency of the averaging approach in analyzing multiscale structures. Reprinted from [15], Copyright 2013, with permission from Elsevier

## 2.3  Multiscale Approaches to Analyze Heterogeneity

Multiscale methods are commonly used in different fields, although most do not use the term 'multiscale.' In fact, both scientists and engineers frequently consider multiscale problems, either intentionally or unintentionally. Multiscale methods can be roughly divided into three types [7, 15]: descriptive, correlative, and variational, as shown in Fig. 2.4.

The most popular is the *descriptive* multiscale method, which is widely used to identify various structures at different scales without, however, accounting for their mechanism of formation, e.g., in image analysis, material characterization, numerical computation, and morphology. Although this method is used mainly for stationary structures, dynamic structures can also be described when statistical properties of structures are needed. Mathematical tools such as wavelet analysis [31] have been used for this purpose.

The *correlative* method tries to establish the relationship between parameters at different scales. This bridges the parameters at element scale with those at system scale by applying coarse graining, constitutive equations, and statistical mechanics. However, it is difficult to establish such a bridge, and the governing rule of the relationship between different scales has not been physically revealed. Accordingly, different approaches are usually based on different assumptions, which can introduce errors. Although much effort has been expended in this kind of analysis, it is not well understood yet.

Most current literature on this topic is related to descriptive and correlative multiscale methodologies.

To correlate different scales that contribute to the formation of multiscale structures, the *variational* multiscale methodology follows a different strategy to those described above. It stipulates that (1) multiscale structure arises from the stability condition of a given system [5, 6, 19], and (2) the phenomena at different scales are determined by this stability condition. Therefore, in addition to structural resolution at each scale, stability conditions are established to reflect the correlation between scales, which involves mesoscale modeling.

Different disciplines at different levels (cf. Fig. 6.3) deal with multiscale problems in various ways, but mostly with coarse graining or

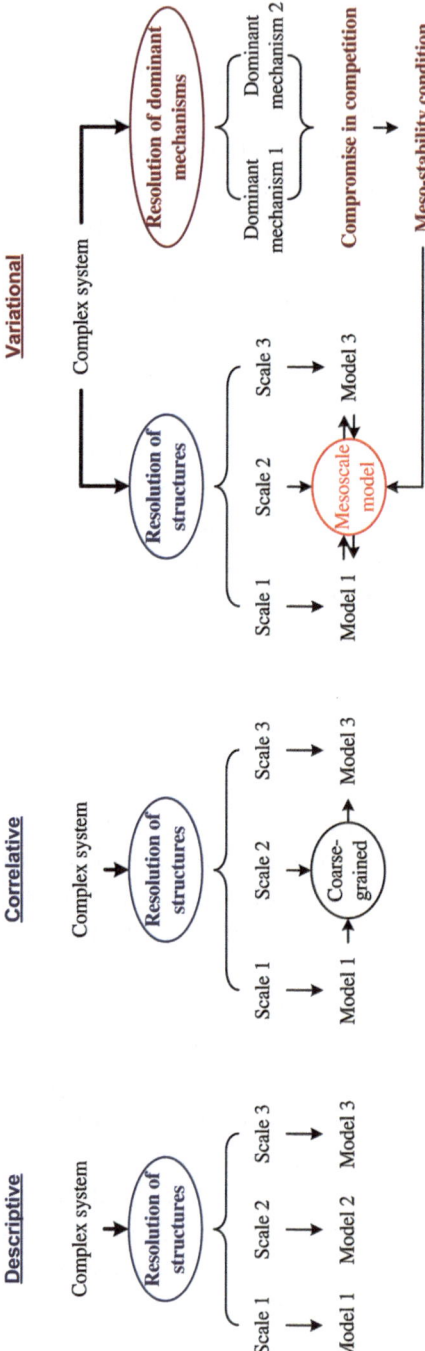

**Fig. 2.4** Multiscale methodologies: descriptive, correlative, and variational

statistical approaches, that is, an averaging approach, which is because of the lack and/or ignorance of the governing principle that rules phenomena at mesoscales. Variational approaches aim to overcome this problem by identifying the stability conditions of structures at these mesoscales to correlate parameters at microscales and macroscales. In fact, a stability condition must be needed when structures are studied, without which the formation mechanism of structures cannot be revealed, and the correlation between parameters at different scales cannot be established.

The variational multiscale methodology also recognizes the multiplicity of multiscale structures, however, because it is difficult to formulate stability conditions, variational multiscale methodology has not yet been sufficiently explored, as reviewed in [7, 12].

Both the correlative and the variational methods can be of help to coarse-grained models and constitutive equations, but in different ways: The former is through graining microscale interactions, which is a unidirectional correlation, and has been applied widely; The latter is through providing with a governing rule or a stability criterion in a bidirectional correlation, deserving to be explored in the future, as discussed in this brief.

## 2.4  A Variational Multiscale Method: The EMMS Model

It is obvious that correlative and variational methods are devoted to establishing mesoscale models but from different angles: the variational method uses a stability condition and the correlative one grains element behaviors. The EMMS model [15] is a variational method that originated from the analysis of particle clustering in gas–solid fluidization, and is a typical mesoscale model. As illustrated in Fig. 2.5, a gas–solid fluidization system contains a heterogeneous structure, roughly consisting of dense and dilute phases, with an average voidage $\varepsilon$ at the global scale under the operating conditions $U_g$ and $U_p$. At the particle scale, individual particles interact with the gas with different parameters in the dense and dilute phases (here, we assumed uniformity in both dense and dilute phases). At the mesoscale between the particle and global scales,

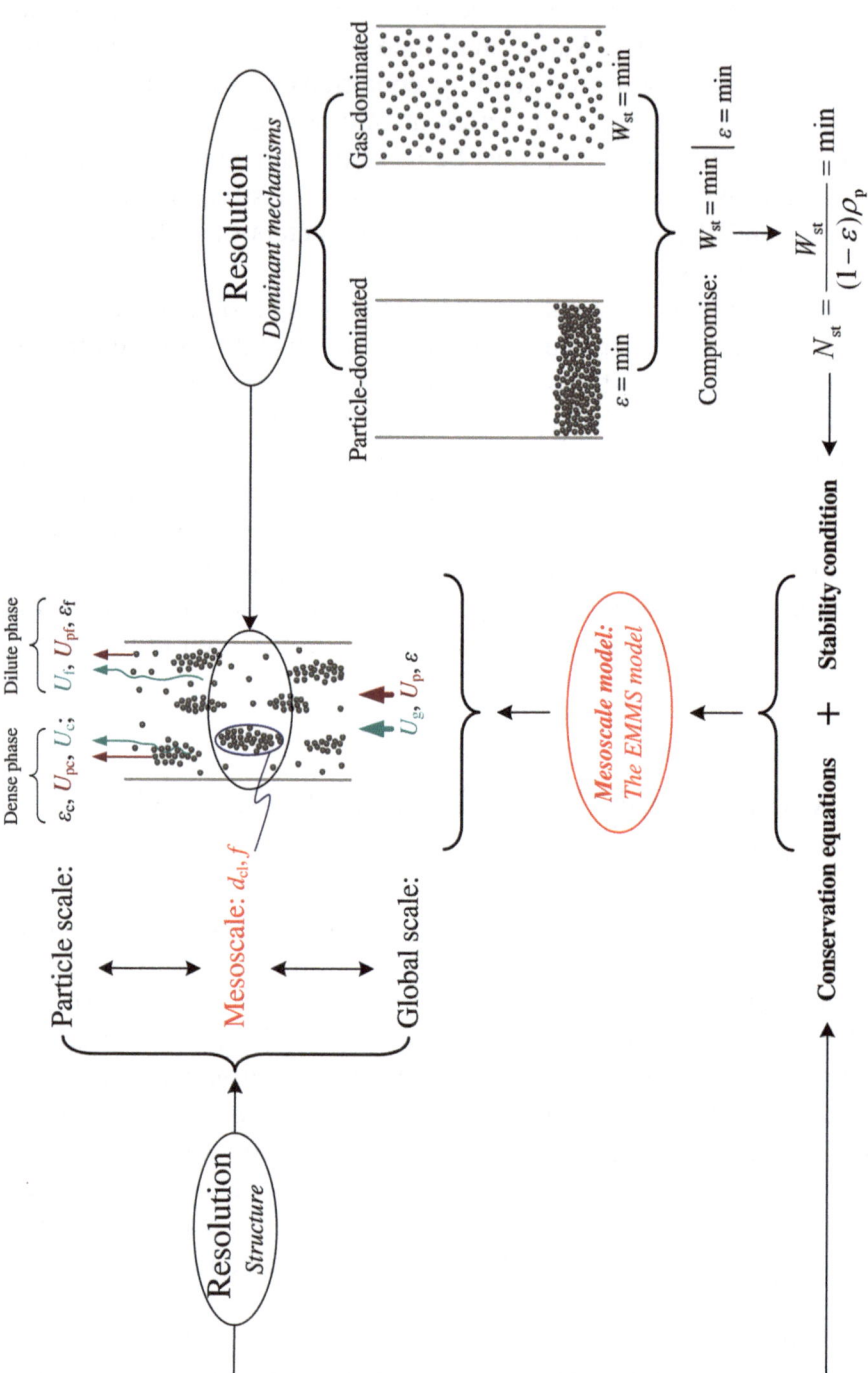

**Fig. 2.5** Framework of the EMMS model. Modified from [50], Copyright 2014, with permission from AIP Publishing LLC

complicated dynamic particle clusters form; these represent the collective behavior of particles. These dynamic clusters complicate the modeling of gas–solid fluidization systems.

Correlating the global parameters $U_g$, $U_p$, and $\varepsilon$ with the particle-scale parameters $\varepsilon_f$, $U_{pf}$, and $U_f$ for the dilute phase and $\varepsilon_c$, $U_c$, and $U_{pc}$ for the dense phase is a challenge. We therefore included mesoscale parameters $f$ and $d_{cl}$, which call for an additional condition besides conservation equations since the number of variables is larger than the number of equations available. While looking for this additional condition, we recognized that the resolution of dominant mechanisms had to be analyzed in addition to structure resolution. The compromise between the dominant mechanism of particles, $\varepsilon = \min$, and that of the gas, $W_{st} = \min$, defines the stability condition of the system, and correlates the parameters at different scales. That is, the energy consumption for suspending and transporting particles per unit mass $N_{st} = \min$, leading to the EMMS model (Fig. 2.5).

From the EMMS model, eight parameters ($\varepsilon_f$, $U_{pf}$, $U_f$, $\varepsilon_c$, $U_c$, $U_{pc}$, $f$ and $d_{cl}$) at three scales can be solved, and regime transitions with totally different transport behaviors can be defined [19]. In particular, the EMMS model can predict the phenomenon of choking, which is a jump change in structure, from operating conditions and material properties [19, 21].

This is the first version of the EMMS model [5, 6, 19]. At the beginning of this work, we used the term "coordination" and $N_{st} = \min$ to describe the compromise in competition between gas and solids because the extremum behavior of gas $W_{st} = \min$ and that of solids $\varepsilon = \min$ were not yet well understood (in Chinese, "coordination" means action of compromise in competition). After establishing the EMMS model to describe mesoscale structure, it was extended to calculate the radial distribution of heterogeneity assuming a global stability condition (the integrated average of $N_{st}$ over the cross-section) [19, 20]. The EMMS model was able to explain why the so-called core-annulus structure forms in fluidized beds. Then, we examined the concept of "compromise in competition" further to elucidate the mechanism of choking, which occurs under critical conditions as a sudden change in structure from a dilute uniform state to a dense two-phase state [28, 29].

With the recognition of the gas movement tendency of $W_{st} = \min$ and that of solids of $\varepsilon = \min$, the concept of compromise in competition between gas and solids was further clarified into three regimes [28, 29, 60]: gas-dominated with $W_{st}$ = min, gas–solid compromising with $W_{st}$ = min$|_{\varepsilon\ =\ min}$, and solid-dominated with $\varepsilon$ = min. Thereafter, we focused on the compromise regime, defined as the *"compromise in competition between different dominant mechanisms,"* that is, the compromise between $W_{st}$ = min and $\varepsilon$ = min, which we then termed the EMMS principle. Here, an extremum represents "competition," and "coordination" implies the responding action in the compromise in competition.

In fact, the mechanism of the stability condition should be the key to solving problems involving dynamic structures. However, it was neglected as researchers looked for alternative solutions such as coarse graining, constitutive, and statistical models for describing dynamic systems.

In a subsequent study, we further confirmed the principle of compromise in competition [23]. In particular, the universality of the EMMS principle was first explored by extending it to single-phase turbulent pipe flow [11], which we described as the compromise in competition between viscous and inertial effects both formulated as extrema. We then made a daring prediction of its universality with respect to all complex systems [28, 29] based on these two successes in gas–solid and single-phase systems. The ability of the EMMS principle to describe multiscale systems promoted us to search for a general principle for all complex systems, that is, mesoscience, even as early as the 1990s [19, 28, 29], although without using such a term. However, we first had to convince the research community of the validity of the EMMS principle, which was a challenge. We spent many years validating the EMMS principle, which we did by verifying its principle, extending it to other systems, and using it to solve practical problems in both fundamental research and industrial applications.

# Chapter 3
# Verification and Application of the EMMS Principle

## 3.1 Proof of the EMMS Principle

Parallel to the extension of the EMMS principle, as discussed in the next two sections, we verified and applied it to respond to controversy over its rationale. We focused on understanding the variations of $N_{st}$ at different scales by performing discrete simulation of the so-called pseudo-particle model [9, 10], which took us more than 10 years. Before unequivocal verification of the EMMS principle, preliminary confirmation of the importance of compromise in competition was demonstrated by the interaction between two streams of discrete media, which form structures. This first demonstration was published in 1998 [23], as described below.

### 3.1.1 The Importance of Compromise in Competition

The EMMS principle was first confirmed by a simple computational experiment, which was called graphic simulation at that time. Two dominant mechanisms were represented by two streams of discrete 2D circular particles of the same size, labeled A and B, to represent two media, moving toward each other with the same velocity, as depicted in Fig. 3.1. Their relative momentum could be adjusted by changing the mass ratio (or equivalently the density ratio) of A to B, $R_m = m_A/m_B$, and the dynamics of either stream could be adjusted independently by changing the ratio of the fluctuating random

J. Li and W. Huang, *Towards Mesoscience*, SpringerBriefs in Applied Sciences and Technology, DOI: 10.1007/978-3-642-41790-0_3,
© The Author(s) 2014

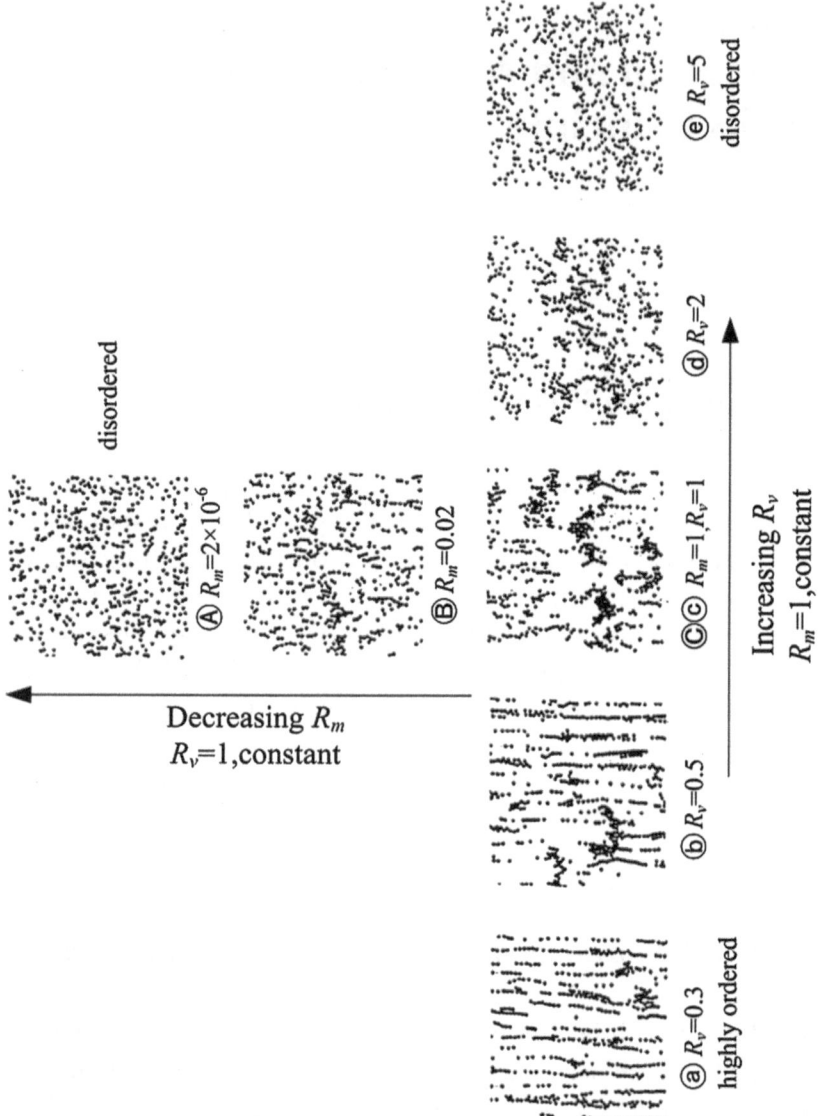

**Fig. 3.1** Computer-aided experiments showing the dependence of flow structure on the relative dominance between two particulate media. **a–e** show changes associated with increasing $R_v$, representing increased independent movement, which lead to less-ordered behavior and uniform structure. **A–C** show changes caused by increasing $R_m$, which characterize the relative dominance of one medium over another and lead to increased heterogeneity with a maximum at $R_m = 1$. Reprinted from [23], Copyright 2013, with permission from Elsevier

velocity to the directed velocity $R_v = v/V$. Here, $R_m$ represents the relative dominance between these two media because both streams move at the same velocity. Increasing $R_m$ corresponds to increasing the dominance of A over B, which changes their interaction. B dominates A when $R_m \ll 1$, whereas compromise between A and B is observed when $R_m = 1$. $R_v$ characterizes the extent to which the streams exhibit random behavior; increasing $R_v$ intensifies independent movement, which suppresses compromise between A and B.

Figure 3.1(C)/(c) displays a snapshot of two interacting particle flows moving in opposite directions with the same momentum ($R_m = 1$) and random velocity fluctuation ($R_v = 1$). An ordered heterogeneous dissipative structure is generated under these conditions. By decreasing $R_m$ from unity to increase the dominance of B over A (that is, changing the extent of their mutual compromise), the flow structure becomes increasingly homogeneous and disordered, as shown in Fig. 3.1(B) and (A). If we suppress the independent movement of A and B by decreasing $R_v$, the flow structure becomes more ordered, as indicated in Fig. 3.1(b) and (a). This change corresponds to the independent movement of particles being suppressed by external forces such as magnetic force. In contrast, increasing $R_v$ increases random fluctuation, magnifying the disordered movement and leading to a more uniform structure, as observed in Fig. 3.1(d) and (e).

The results from these simulations indicate that the relative dominance between the two media plays an important role in shaping the dissipative behavior in such systems. The exclusive dominance of one over the other destroys such dissipative behavior. The ordered behavior in the system is therefore attributed to the compromise in competition between the two media.

It is clear that a dynamic system containing a mixture of ordered and disordered behaviors is generated through the compromise between two streams attempting to realize their respective movement tendencies (i.e., competition). These results gave us confidence to extend the principles of the EMMS model to produce a general principle for complex systems. This extension of the EMMS model was later confirmed in simulations of two interacting streams of granular flows [12]. A stream of granules move under an external force toward the opposite boundary if another stream is not present, as illustrated in Fig. 3.2. That is, the movement tendencies of the two

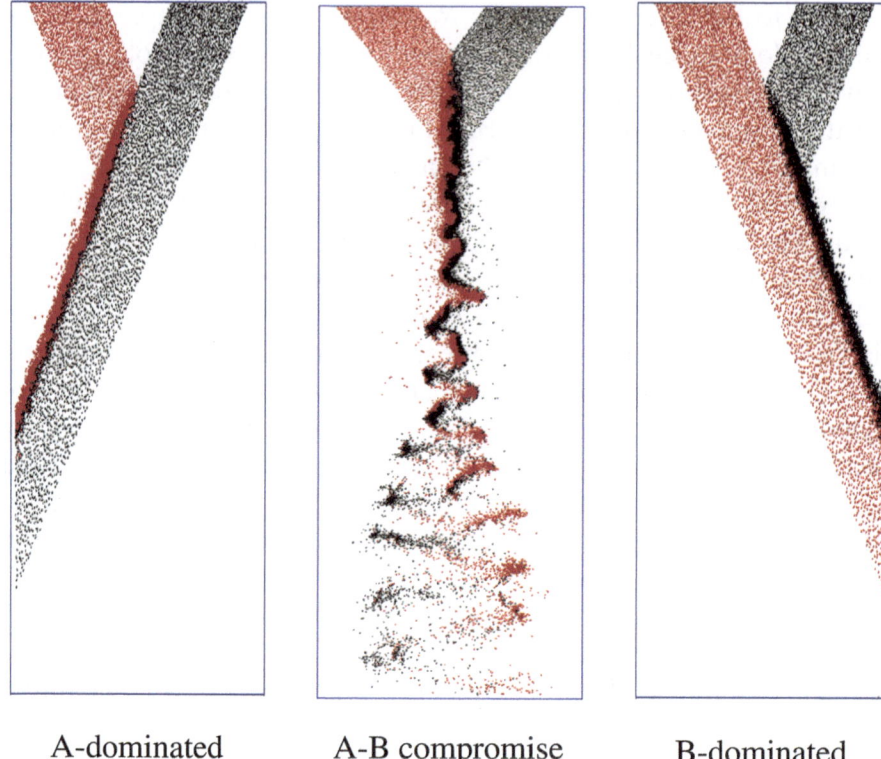

A-dominated          A-B compromise          B-dominated

**Fig. 3.2** Two streams of granular flow: One-stream dominated (*left* or *right*), and two-stream compromise (*middle*) cases. Particle mass is same for both streams. Modified from [12], Copyright 2013, with permission from Elsevier

streams can be described by the minimization of the particle potential in the respective external force fields; intuitively, they are identified as the dominant mechanisms in the system. Similar to gravitational potential energy, the potential energy of particles in their own external force fields $F$, $H_A$, or $H_B$, is proportional to its distance to the destination boundary. By assuming the potential energy at the left (or right) boundary is equal to zero for particles in stream A (or B), the potential energy of particles is $H_i = m (F_i \cos\alpha) L_i$, $i = $ A, B, where $m$ is particle mass, $F_i \cos\alpha$ is the horizontal component of external force field $F_i$, and $L_A$ (or $L_B$) is the distance to the left (right) boundary for the particles in stream A (or B). The extremum tendencies of particle movement can be expressed mathematically as $H_A \rightarrow$ min and $H_B \rightarrow$ min for streams A and B, respectively.

Once these streams are in contact, the "compromise" between their dominant mechanisms $H_A \to$ min and $H_B \to$ min begins. The movement of each stream will be resisted to a certain extent by the other stream so that its extremum tendency cannot be realized in the small region where they interact. However, we assume that the movement tendency of a stream can still be observed when a larger region is considered. Once again, a complicated structure is generated only when two streams compromise with each other.

These two studies established our confidence of the importance of compromise in competition between dominant mechanisms in complex systems. This encouraged us to explore other systems to confirm that compromise in competition between dominant mechanisms is indeed a governing rule for dynamic heterogeneous structures, and is a general mesoscale phenomenon for all mesoscales.

### 3.1.2 Exploring the Universality of the EMMS Principle

The above examples reveal that the stability condition of complex systems can be defined by the compromise in competition between dominant mechanisms, and the prevailing structures of complex systems are likely characterized by a combination of states corresponding to respective dominant mechanisms. That is, a system dominated by two mechanisms shows alternating features with respect to space and time between the state dominated by one mechanism and that by the other. Although the dominant mechanisms differ from system to system, they follow the same rule of compromise in competition to shape the resulting structures. This is a very important clue that linearity (simplicity) and nonlinearity (complexity) can be unified by the principle of compromise in competition. In fact, compromise in competition is ubiquitous at mesoscales, leading to the complexity, diversity, and dynamics of different systems. Figure 3.3 outlines the mechanism of the EMMS principle using gas–solid flow as an example.

On the one hand, compromise *in space* leads to the appearance of a two-phase structure consisting of a particle-rich dense state dominated by particles ($\varepsilon =$ min) and a gas-rich dilute state dominated by gas ($W_{st} =$ min) to reach $N_{st} =$ min at the mesoscale. That is,

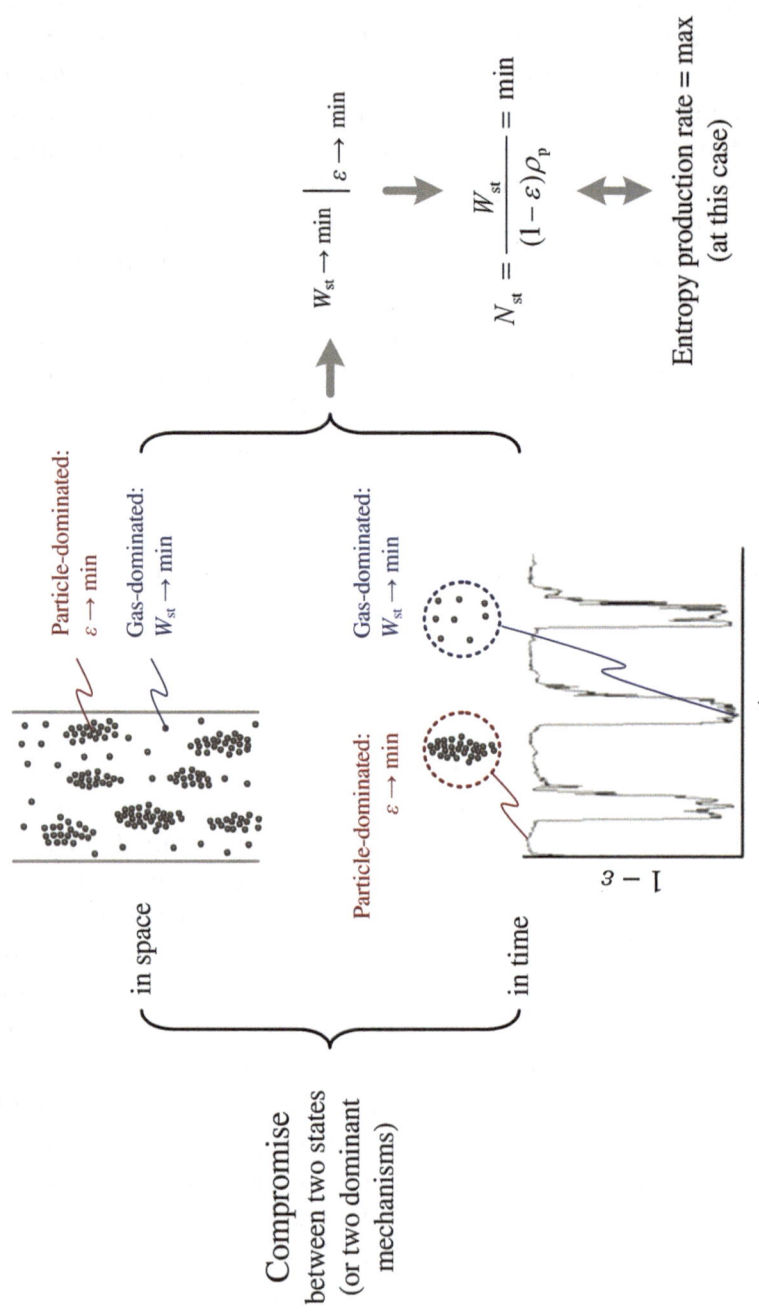

**Fig. 3.3** Illustration of compromise in competition between two dominant mechanisms. Reprinted from [15], Copyright 2013, with permission from Elsevier

when a point is dominated by $\varepsilon = \min$, its neighboring point is dominated by $W_{st} = \min$, and vice versa. On the other hand, compromise *in time* leads to dynamic change at any location; that is, the alternate appearance of particle- and gas-dominated states. With increasing dominance of gas over particles by increasing gas velocity, the fraction of time dominated by gas increases. It is the compromise in competition between gas and solids in both space and time that leads to the stability of the heterogeneous structure, which, in the thermodynamic sense, corresponds to a maximum entropy production rate for the case of gas–solid two-phase flow. All of the systems we have analyzed follow the same principle of compromise in competition to achieve steady heterogeneous structures.

We therefore proposed the following equation as a common variational criterion to describe systems dominated by two mechanisms [28, 29]:

$$\text{Extremum tendency 1} \big|_{\text{Extremum tendency 2}} \cdot \tag{3.1}$$

This is the EMMS principle: To reveal the formation mechanism of a multiscale structure, the individual movement tendency for each dominant mechanism and the compromise in competition between these dominant mechanisms have to be analyzed first. In Eq. (3.1), the extremum tendency implies "competition," and mutual constraint between two extremum tendencies implies "compromise." To realize compromise in competition, coordination of two mechanisms (either inter- or intra-) is needed, which is the principle of compromise in competition; that is, the EMMS principle.

What is worth exploring further is the relationship between the number of dominant mechanisms prevailing in a complex system and the number of characteristic scales in its structure. Whether such a correlation exists needs to be confirmed [29]. Mathematical tools are used to formulate the compromise in competition between dominant mechanisms, which usually involves a MOV problem, and is still a challenge for mathematicians.

Importantly, this principle of compromise in competition is common to all mesoscales, in particular, to different levels of chemical science [13, 14, 18]. Chemical processes cover multiple levels, including chemistry, chemical engineering, and process system engineering, as will be discussed in Fig. 6.3, and each level also possesses a multiscale structure. The chemistry level consists of three scales: atoms or molecules, atomic clusters or molecular assemblies, and bulk materials (e.g., particles, tubes, and films). Chemical engineering includes the scales of particles, particle clusters, and reactors, while process system engineering covers single reactors, factories, and ecological systems. Atomic clusters or molecular assemblies, particle clusters, and factories are the mesoscales of the levels of chemistry, chemical engineering, and process system engineering, respectively. Although these mesoscales involve different structures, phenomena, and performance, they show a common feature: ***Compromise in competition between dominant mechanisms in physics and MOV in mathematics***. This is a very important point for the future of research in chemical engineering [13].

## 3.2  Verification of the EMMS Principle

The above discussion is qualitative and not sufficiently convincing. Therefore, complete verification of $N_{st}$ = min was carried out and finished later [7, 8], which completely confirmed the generality of the EMMS principle.

Based on the simulation of the pseudo-particle model [9, 10], Fig. 3.4 presents the computational and statistical results determined for the variations of $N_{st}$ at different scales with respect to space and time to depict the dynamic process of compromise between two competing mechanisms ($W_{st}$ = min and $\varepsilon$ = min). The system reaches a steady state through the compromise in space and time between $W_{st}$ = min and $\varepsilon$ = min, both of which represent a simple law ($W_{st}$ = min for gas dominating, and $\varepsilon$ = min for solids dominating). The mechanism of the compromise process [7, 8] is discussed below.

At the microscale (for instance, at any given point A in Fig. 3.4), the two extremum tendencies, $W_{st}$ = min and $\varepsilon$ = min, can be realized only alternately over time. With respect to any given time, when $W_{st}$ = min occurs at point A, $\varepsilon$ = min would occur elsewhere at another nearby point, such as B. This implies that linear or simple laws could play roles in nonlinear complex systems but only locally at a given point or temporarily at a given instant, indicating that there is no stability condition at the microscale.

In a sufficiently large region, that is, the mesoscale designated by D, $W_{st}$ = min and $\varepsilon$ = min cannot be realized simultaneously

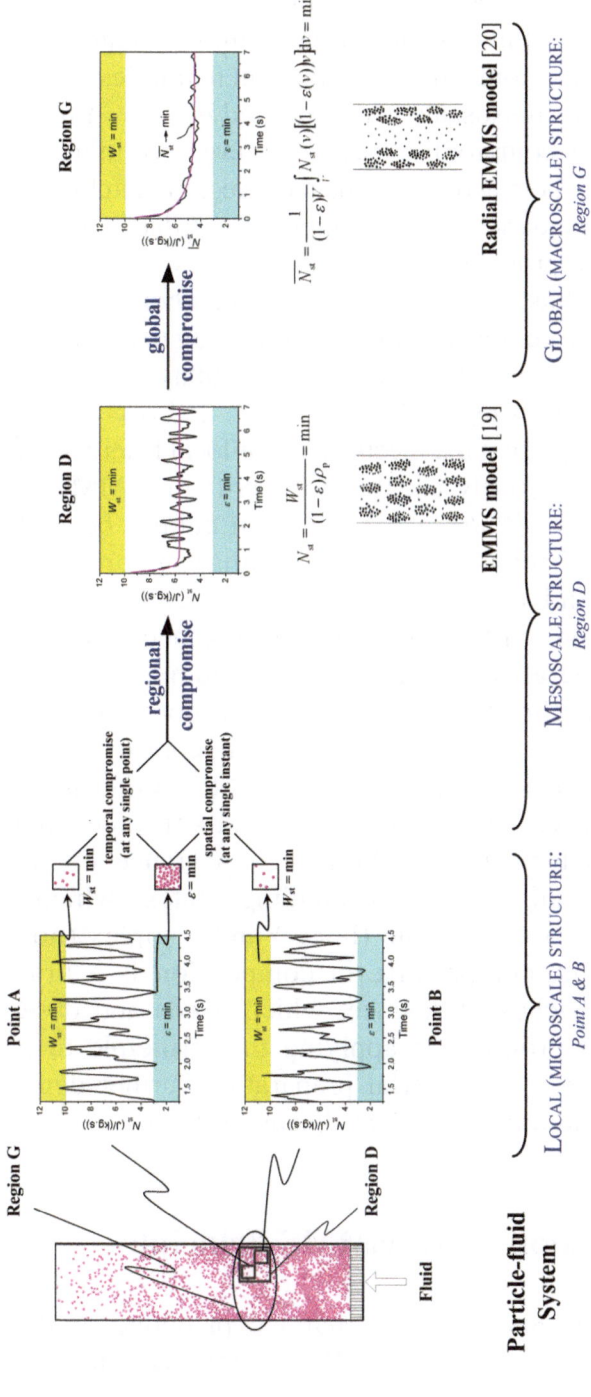

**Fig. 3.4** Verification of the EMMS principle. Modified from [12], Copyright 2013, with permission from Elsevier

even for an instant. However, spatio-temporal compromise between the particles and fluid occurs in the volume, leading to an averaged $N_{st} = $ min with some fluctuation, which implies that the stability condition at this mesoscale is represented by $N_{st} = $ min.

Globally, this compromise leads to an averaged minimum $N_{st} = $ min integrated over the macroscale labeled as region G in Fig. 3.4. The effect of the wall boundaries on the compromise reduces $N_{st} = $ min further, that is, the wall promotes $\varepsilon = $ min to occur more frequently at the wall region, but inhibits $W_{st} = $ min. This causes particles to move to the wall, and the gas to accumulate at the center, leading to a radial distribution in the system. So, $N_{st} = $ min represents the stability condition at the system scale. $N_{st} = $ min in region D (mesoscale) and $N_{st} = $ min in region G (macroscale) represent the mechanisms of local heterogeneity and radial distribution, respectively, and therefore verify the EMMS principle.

In fact, $N_{st} = W_{st}/((1-\varepsilon)\rho_p) = $ min cannot be realized at any specific point because of the compromise in competition between $W_{st} = $ min and $\varepsilon = $ min. At the mesoscale, their compromise in competition with respect to both space and time leads to $N_{st} = $ min overall in the mesoscale volume. That is, the EMMS principle is applicable to mesoscales but not to microscales.

This progress removed the initial controversy surrounding the EMMS principle, and promoted us to extend it further and to explore its universality. The most important understanding gained from this analysis is the alternate appearance of two dominant mechanisms and their corresponding states, which reveals the relationship between simple (linear) and complex (nonlinear) processes. That is, compromise in competition is a bridge between these two branches of science, which needs and deserves further study. Recent progress in applying this principle to protein folding, turbulence, and material preparation further demonstrated its importance and universality, as discussed later.

## 3.3 Application of the EMMS Principle

Application is another way that we provided evidence for the EMMS principle. The EMMS model was first successfully applied in 1998, when we used it with SINOPEC to scale-up a fluid cracking catalytic process directly from laboratory to commercial scale [26]. We also used the EMMS model to predict the drag between gas and solids in heterogeneous structures [19]. However, we did not apply the EMMS drag model to computational fluid dynamics (CFD) until 2003 [33] because of mathematical difficulties in solving the EMMS model. When we

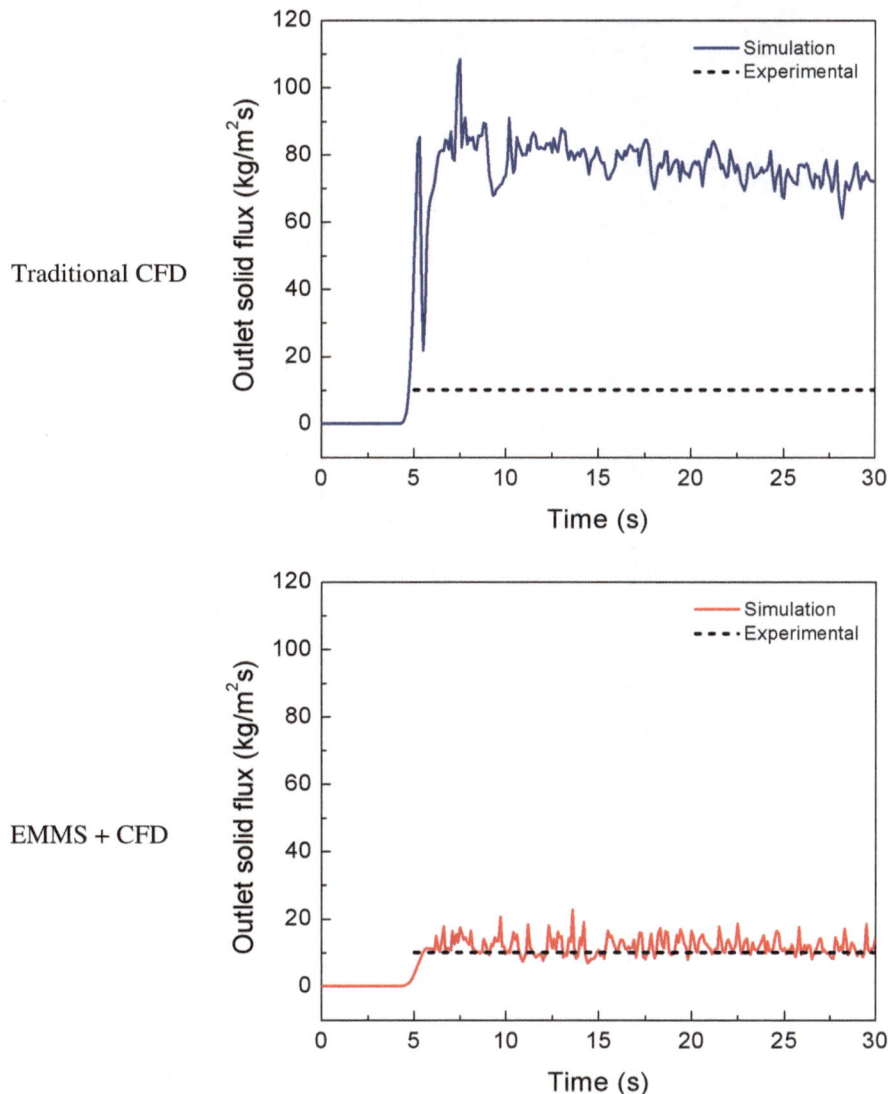

**Fig. 3.5** Comparison between traditional CFD and EMMS-CFD using a global average of the EMMS drag. Modified from [33], Copyright 2014, with permission from Elsevier

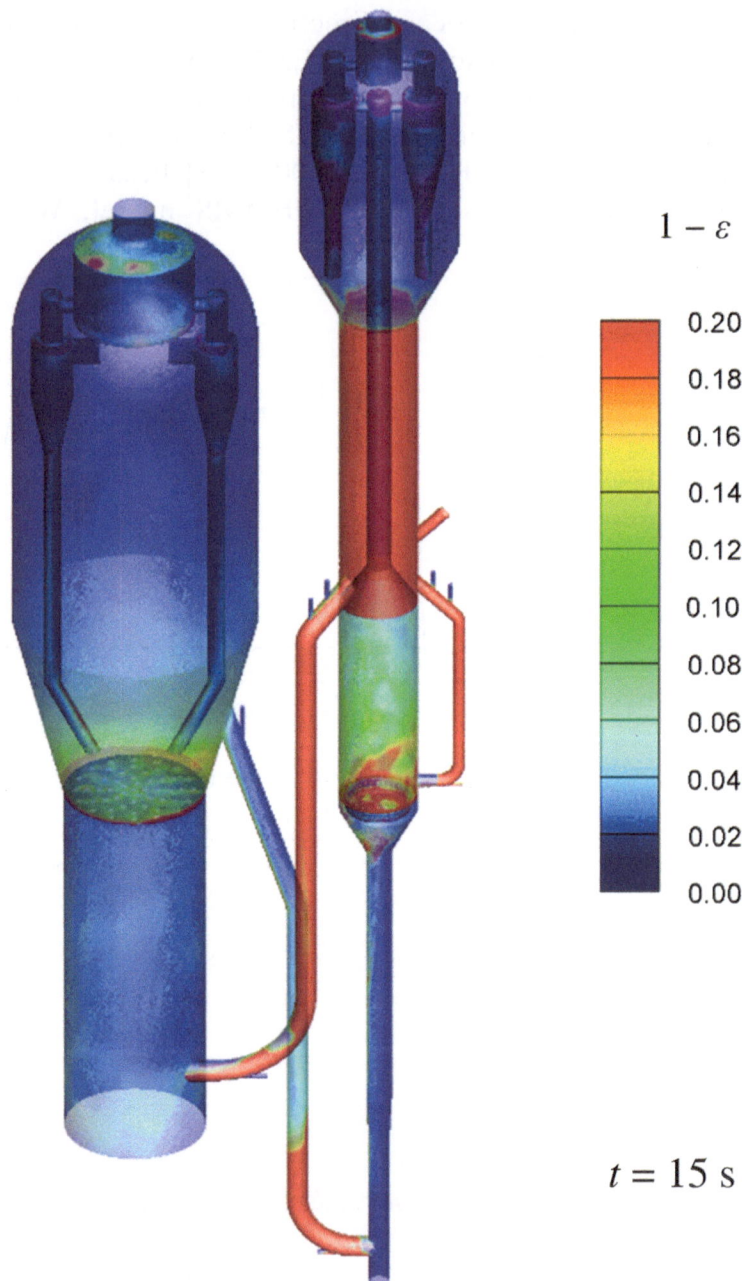

$1 - \varepsilon$

0.20
0.18
0.16
0.14
0.12
0.10
0.08
0.06
0.04
0.02
0.00

$t = 15 \text{ s}$

**Fig. 3.6** Simulation of 1.4 Mt/a fluid cracking catalytic reactor with EMMS drag applied in subgrids. Modified from [18], Copyright 2014, with permission from Elsevier

recognized this problem, efforts were made to find a simple solution for the model [26, 61]. In 2002, a very simple solution was realized [32], enabling us to use the EMMS drag model in CFD. Applying the EMMS drag improved the predictability of commercially available CFD codes, and it was first successfully applied to CFD simulation with a global average in 2003 [33], achieving good agreement between simulation and experiment data, as presented in Fig. 3.5.

We then extended the EMMS drag model to the subgrid scale of CFD [37], simulating the hydrodynamics of a whole loop of a circulating fluidized bed (CFB) reactor with a height of 50 m and diameter of 1–3 m, as shown in Fig. 3.6. Such a simulation is impossible without considering the mesoscale structure and the choking behavior of the system. This example revealed the great potential of mesoscale modeling based on the EMMS principle to replace coarse graining modeling in CFD. The success of this simulation is attributed to the good predictability obtained from the mesoscale model and the scalability arising from the ability to use much larger grids.

These efforts to verify and apply the EMMS principle gave us confidence to explore the universality of the EMMS principle by extending it to other systems, and to look for the possibility to realize VPE through mesoscale modeling.

# Chapter 4
# Extension and Generalization of the EMMS Principle

## 4.1 Extension of the EMMS Principle to Gas–Liquid Systems

Parallel to the verification and application of the EMMS principle, and along with the preliminary indication of its possible generality from its extension to turbulent flow [11], we extended the EMMS principle to emulsions, granular flow, foam drainage, gas–liquid flow, and nanoscale gas–liquid flow. All of these investigations gave promising results [12] indicating the generality of the EMMS principle.

The EMMS principle was extended to gas–liquid systems, as shown in Fig. 4.1 (which is similar to Fig. 2.5 for the original EMMS model for a gas–solid system). The compromise in competition between the liquid-dominated $N_{surf}$ = min and the gas-dominated $N_{turb}$ = min allows us to correlate parameters at different scales. Importantly, the sudden change in structure observed in such systems can be predicted by the model [38].

## 4.2 Extension of the EMMS Principle to Turbulence

Recently, on the basis of previous studies [11, 12], the EMMS principle was further used to close the turbulence model by considering the compromise in competition between viscosity-dominated $W_v$ = min and inertia-dominated $W_{te}$ = max, as illustrated in Fig. 4.2. Reasonable agreement was obtained between simulation and experimental data [49].

J. Li and W. Huang, *Towards Mesoscience*, SpringerBriefs in Applied Sciences and Technology, DOI: 10.1007/978-3-642-41790-0_4,

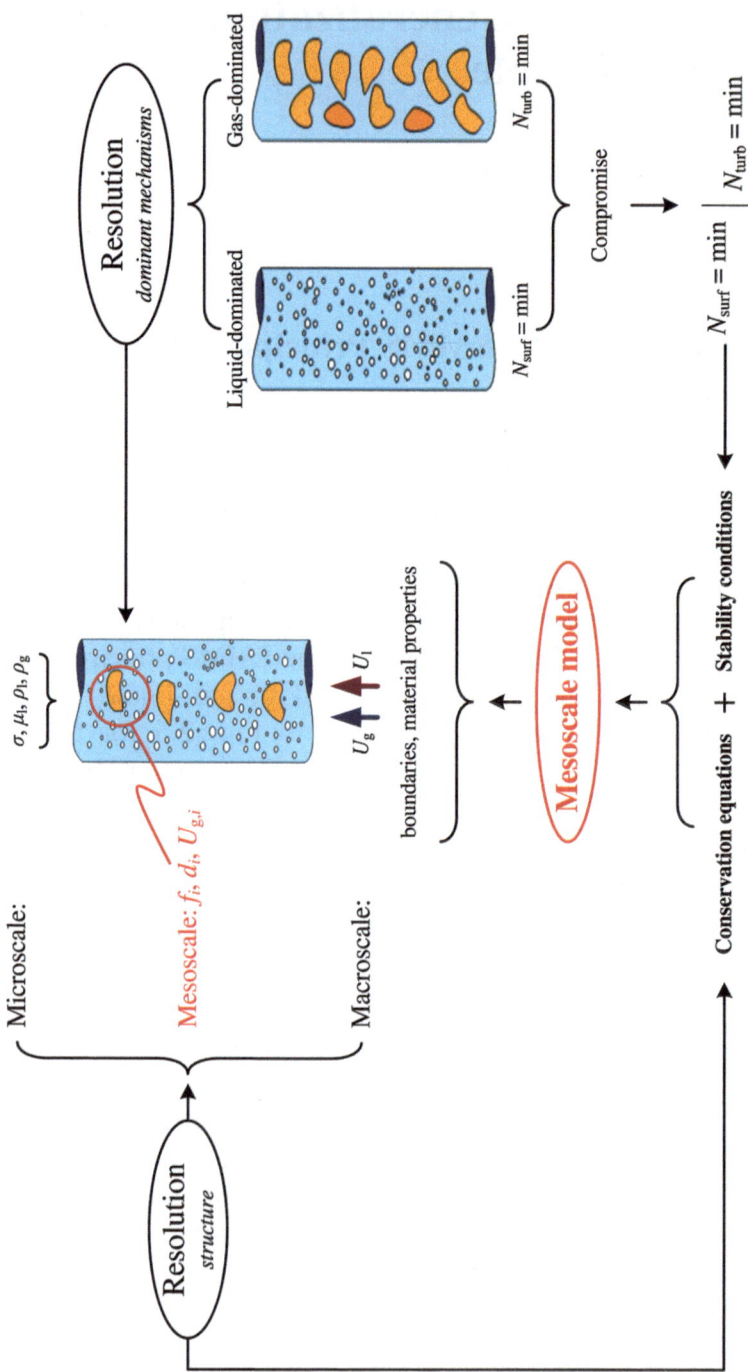

**Fig. 4.1** A gas–liquid system according to the EMMS model

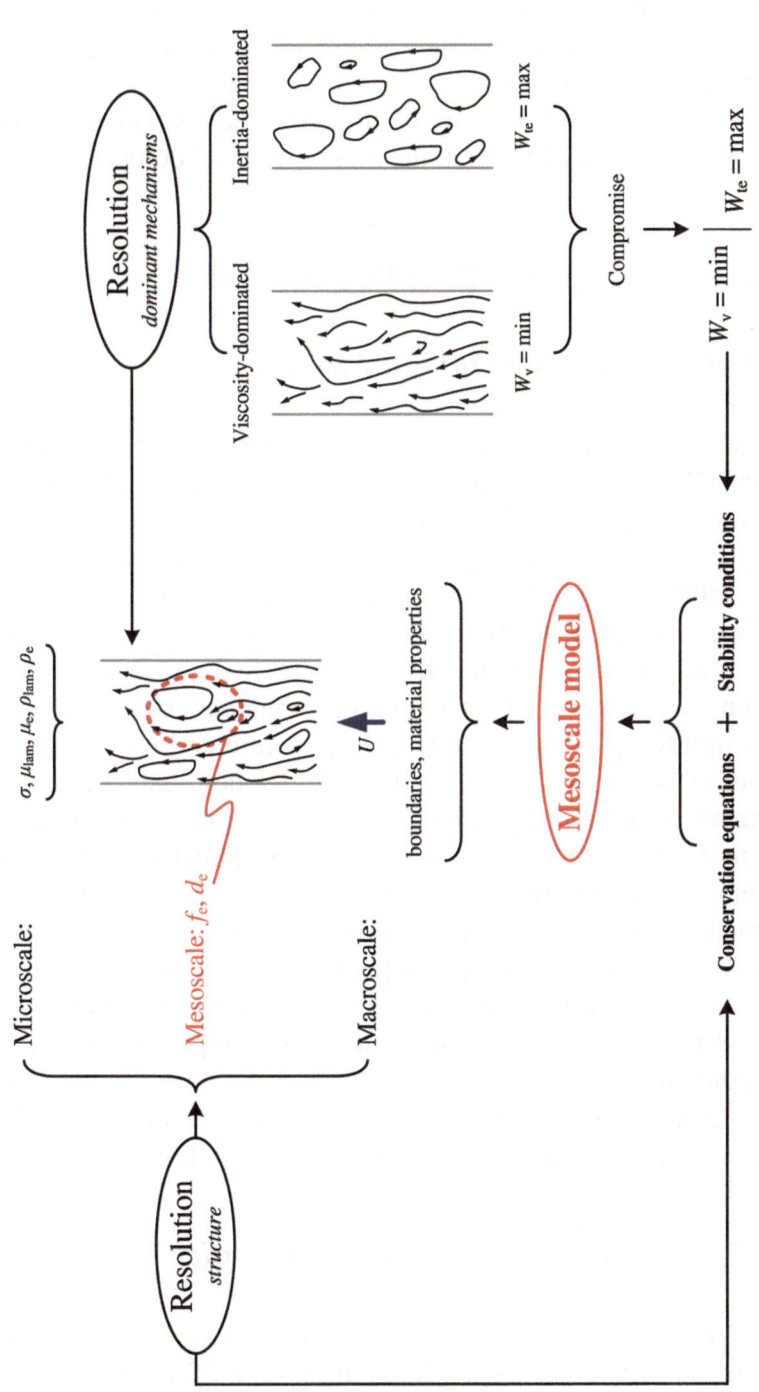

**Fig. 4.2** EMMS turbulence model [49]

The universality of the principle of compromise in competition between dominant mechanisms was further confirmed by applying it to other systems [12]. Our results allowed us to preliminarily conclude that the compromise in competition between dominant mechanisms is the origin of complexity and diversity of all mesoscale phenomena and represents the mechanism of the stability condition of all complex systems.

## 4.3   Universality of the EMMS Principle

### 4.3.1   The EMMS Principle in Physics

After studying seven different systems and exploring the universality of the EMMS principle, as summarized in Fig. 4.3, it is clear that stability conditions play a central role in establishing multiscale models for complex systems. Although the concrete mathematical expressions of these stability conditions vary from system to system, the physical mechanism leading to them is essentially identical. That is, in each system, multiple dominant mechanisms are first presented, each of which can be expressed as an extremum tendency, e.g., $W_{st} \to$ min and $\varepsilon \to$ min for gas–solid systems, $N_{surf} \to$ min and $N_{turb} \to$ min for gas–liquid systems, and so on [12]. Different dominant mechanisms compromise in their competition with one another in a coordinated fashion to form complex multiscale dynamic structures. The stability condition is the dominant rule of such compromise, which shapes the characteristics of mesoscale structures, and defines the correlation between scales. Compromise in competition leads to heterogeneity and dynamic changes, while coordination results in self-organization.

The common principle of compromise in competition will be the unified core of mesoscience. Understanding this common principle requires further effort, particularly with respect to all of the dominant mechanisms existing in different complex systems.

### 4.3.2   The EMMS Principle in Mathematics

Mathematically, the principle of compromise in competition can be expressed as a MOV problem. The stability condition allows closure

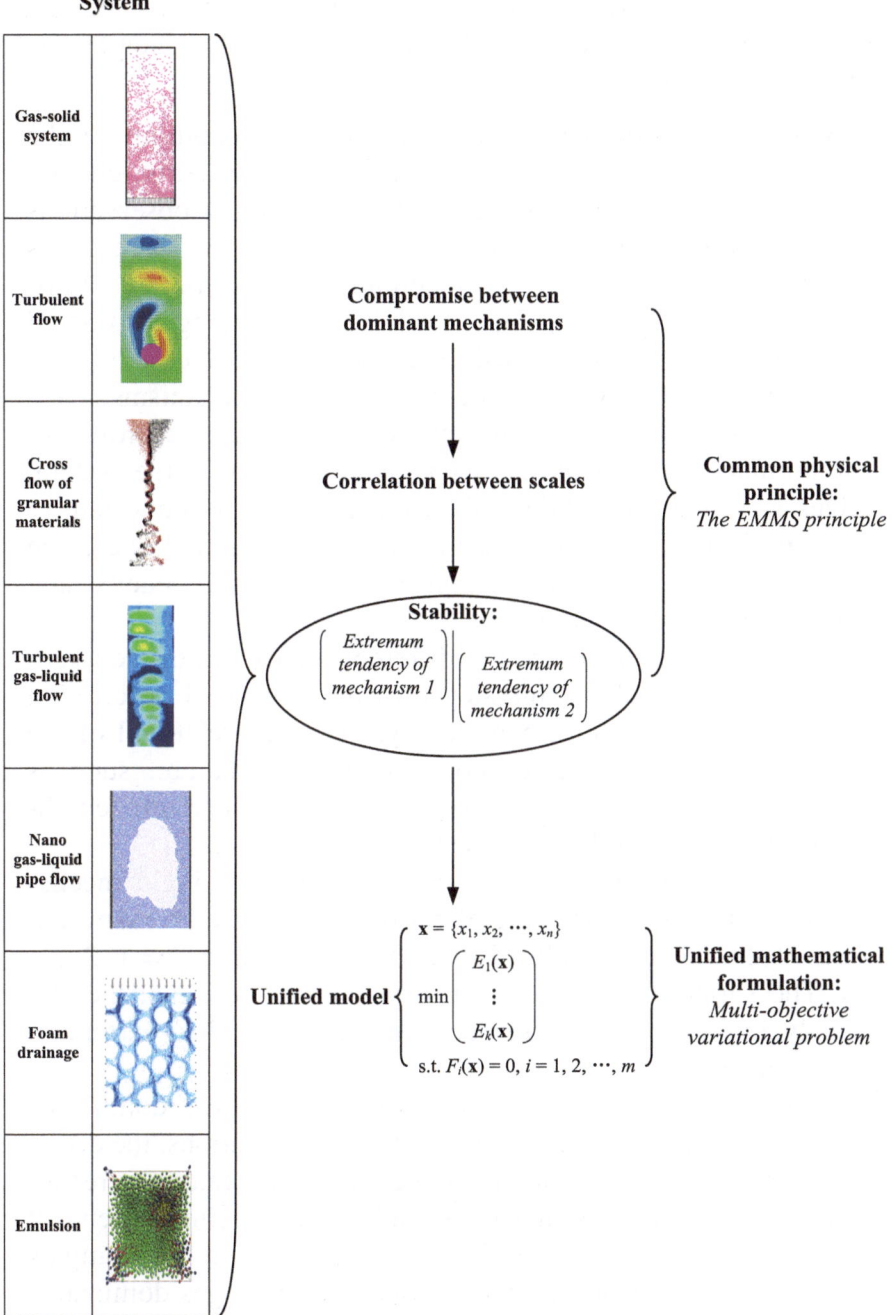

**Fig. 4.3** Common physical principle and general mathematical formulation for different systems that indicate the possible universality of the EMMS principle. Modified from [18], Copyright 2014, with permission from Elsevier

of the variables in the equation set that defines the multiscale model for a specific system. The formulation of the EMMS principle can be summarized as follows:

1. System resolution with respect to scales: According to the characteristic multiscale nature of complex systems, characteristic scales should first be identified reflecting observations of structure or inflective changes. Scale-specific parameters should then be assigned to quantitatively describe the multiscale structure, $\mathbf{x} = \{x_1, x_2, ..., x_n\}$. The level being considered must be specified first, as will be emphasized in Sect. 6.6.
2. System resolution with respect to dominant mechanisms: Each of $k$ dominant mechanisms involved is ranked according to importance to discover the mechanisms that govern the system. This step is critical though difficult. Likely, it is important to group all dominant mechanisms into two integrated ones: one corresponding to the interaction promoting heterogeneity, and the other that suppresses it.
3. Describing the phenomena at different scales and the relationship between scales: This step provides a general description of the phenomena in a complex system to correlate all of the parameters and processes controlling the structure, such as mass and momentum conservation. The conservation conditions are formulated as $F_i(\mathbf{x}) = 0, i = 1, 2, ..., m$.
4. Establishing the variational criterion: For any single mechanism in action, without the effect of others, its extremum tendency should be formulated such that $E_j(\mathbf{x}) = \min, j = 1, 2, ...,$ $k$. The variational criterion for a system can be established by analyzing the compromise between these individual extremum tendencies.
5. Regime identification: By evaluating the relative dominance between different mechanisms at specified conditions, the dominant mechanisms in action could be identified, leading to recognition of different regimes of operation. This regime-dependent feature of variational criterion is very important for complex systems. For instance, in gas–solid systems, gas-dominated, gas–solid compromising, and solid-dominated regimes give quite different structures and performance.

6. Integrating the variational criterion and conservation conditions for a system: This step identifies the steady state of the
   system from all possible solutions defined by the conservation
   conditions $F_i(\mathbf{x}) = 0$, $i = 1, 2, \ldots, m$.
7. Mathematical formulation of the EMMS principle:
   Summarizing the above steps, the EMMS principle could be
   mathematically expressed as follows. For mesoscale structures
   in complex systems defined by $n$ variables $\mathbf{x} = \{x_1, x_2, \ldots, x_n\}$
   dominated by $k$ dominant mechanisms, assuming its conserva-
   tion conditions can be formulated by the set of functions of $\mathbf{x}$;
   that is,

$$F_i(\mathbf{x}) = 0, \; i = 1, 2, \ldots, m < n, \tag{4.1}$$

and the extremum tendencies of these dominant mechanisms
are expressed as $E_j(\mathbf{x}) = \min$, $j = 1, 2, \ldots, k$ (note that a maximum can always be expressed mathematically as a minimum
of its negative [3]), the steady state of the system could be formulated for different regimes as

$$\min E_j(\mathbf{x}), \text{ subject to } F_i(\mathbf{x}) = 0, \; i = 1, 2, \ldots, m \tag{4.2}$$

when dominant mechanism $j$ exclusively dominates the system.
When none of the mechanisms dominate, then

$$\min \begin{pmatrix} E_1(\mathbf{x}) \\ E_2(\mathbf{x}) \\ \ldots \\ E_k(\mathbf{x}) \end{pmatrix}, \text{ subject to } F_i(\mathbf{x}) = 0, \; i = 1, 2, \ldots, m. \tag{4.3}$$

In fact, the original EMMS model can be deduced from this generalized formulation by assuming $k = 2$ and $n = 8$.

This MOV problem is still a challenge for mathematicians. Its
general solution is far beyond our capacity as chemical engineers;
however, it is worth making an effort to explore its nature to help
understand complex systems. This formulation defines the relationship between variables at different scales, including intrascale
interaction and interscale communication, so it also defines a computation paradigm that we call *the EMMS paradigm*. Because of its

ability to extract the most essential characteristics of complex systems, it is possible to simplify the computation of complex systems and thereby expand computation capability considerably, as discussed next.

## 4.4 The EMMS Paradigm for Computation

### 4.4.1 Framework of the EMMS Paradigm

The EMMS principle defines the relationship between parameters at different scales (Eq. (4.3)), and also expresses the respective stability conditions at the mesoscale and the macroscale, as verified in Fig. 3.4. That is, at the system scale, computation is governed by the global stability condition, operating conditions and boundary conditions; at the mesoscale, computation is dominated by the mesoscale stability condition and the interaction between grids; at the microscale, computation mainly addresses the interaction between elements. Therefore, the EMMS principle also defines a multiscale paradigm of computation, the EMMS paradigm.

Figure 4.4 reflects this scale-specific nature for computation, data intensity, and communication in a gas–solid system [20], which are likely related to the interaction between elements and grids, number of elements and grids, and the correlation between scales, respectively. At the microscale, particle-fluid interaction was approximated and formulated using a two-phase model; that is, the dense and dilute phases governed by fluid and particle interactions in each phase, respectively. At the mesoscale, the stability condition $N_{st}(r) \to$ min was used to formulate the bulk structural characteristics and the interaction between the two phases in a volume by considering the interaction between volumes (reflected by related experimental data). Finally, at the macroscale, the radial distribution of solid particles was defined by the global stability condition $\overline{N}_{st} \to$ min including boundary and operating conditions; that is, the radial profile [20] was subject to

$$\overline{N}_{st} = \frac{2}{R^2(1 - \bar{\varepsilon})} \int_0^R N_{st}(r)(1 - \varepsilon(r)) r \mathrm{d}r = \min, \qquad (4.4)$$

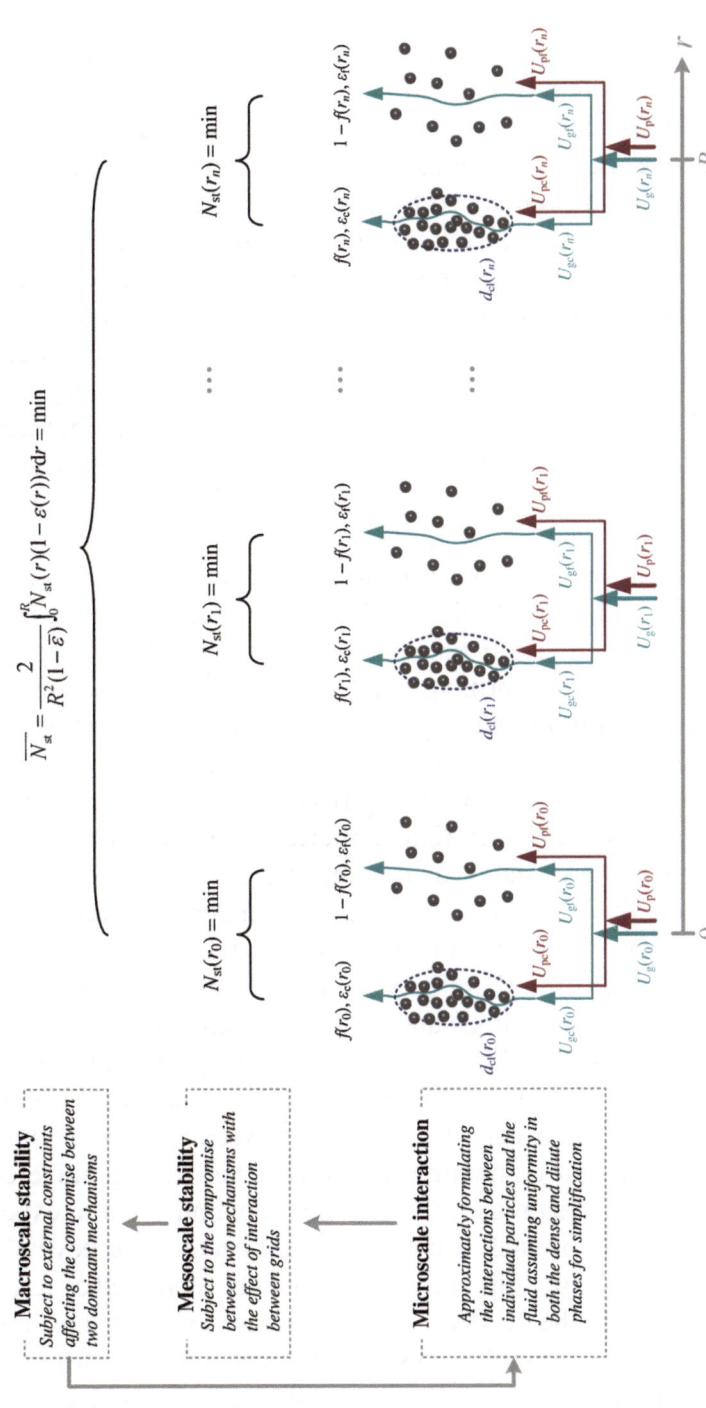

**Fig. 4.4** Three-scale simulation of the hydrodynamics of CFB risers using the EMMS model—the first version of EMMS paradigm. Modified from [18], Copyright 2014, with permission from Elsevier

where $\bar{\varepsilon} = \frac{2}{R^2} \int_0^R \varepsilon(r) r dr$, which yields a reasonable prediction of the radial distribution of particle velocity using experimental data for particle concentration and gas velocity. This was in fact the first version of the EMMS paradigm of computation, and was practiced as early as more than 20 years ago [20]. However, it was tested only for CFB systems at that time. Because of a lack of understanding of its generality, we did not propose a generalized form of this model until 2003 [15].

Figure 4.4 defines the relationship between different parameters at different scales, and presents the stability conditions for the mesoscales and macroscales in a CFB. That is, stability conditions constitute a global effect as a result of the compromise in competition between different dominant mechanisms such that different regimes of systems and their transitions (typically jump changes) can be defined at the macroscale, while the statistical characteristics of structures (typically heterogeneous and dynamic) are defined at the mesoscale. At the microscale, stability conditions cease to function as a whole, and the dominant mechanisms start to appear individually and alternately in space and time. At this scale, the systems display a largely discrete nature where short-range and local interactions dominate. This allowed the EMMS principle to be developed into a multiscale computational paradigm featuring scale-dependent computation, communication and storage defined by physical modeling.

This means that the simulation of physical systems is implemented on different scales using different numerical methods with communication between these scales, as demonstrated in Fig. 4.5. Currently available computers can, of course, be used under this paradigm but the high accuracy, capability, and efficiency (ACE) required by VPE, cannot be achieved using them because of the mismatch between the diversity of numerical methods and rigidity of hardware design.

To further upgrade ACE, a multiscale architecture of hardware systems consistent with the EMMS paradigm is needed to achieve structural similarity between the simulated system, physical model, numerical method, and computing hardware. This will optimize computational capability, storage capacity, and communication expense. Although the details of such a multiscale architecture are beyond our current knowledge, some important features can be identified as follows.

The essence of the EMMS paradigm is to correlate multiscale phenomena in complex systems with stability conditions to reflect the compromise in competition between different dominant

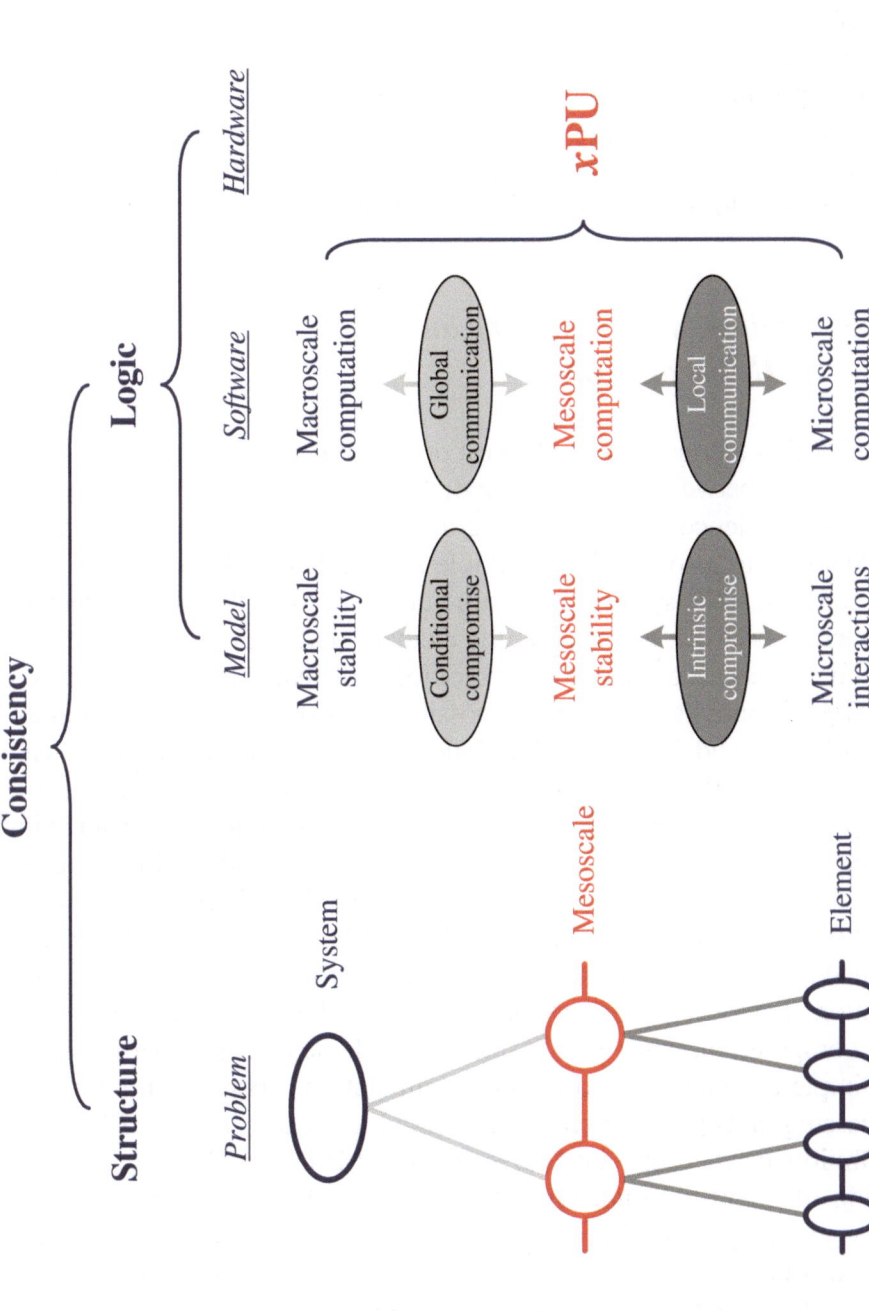

**Fig. 4.5** Overview of the EMMS paradigm of computation. Modified from [3], Copyright 2013, with permission form Springer

mechanisms in a system, and to reasonably characterize mesoscale structures critical for system behavior. Stability conditions at macroscales and mesoscales play dominant roles in controlling the correlations between scales. Considering that the multiscale nature and the compromise in competition between dominant mechanisms are universal for most complex systems in engineering, the EMMS paradigm is general for most problems and, therefore, shows importance and potential to realize VPE with sufficient ACE.

## 4.4.2  Similarity of Structure and Logic in the EMMS Paradigm

Figure 4.5 is a conceptual summary of the EMMS paradigm, featuring consistency between problem, model, software, and hardware. Most systems in science and engineering are of a multiscale nature, so their models should also consider multiple scales. This will mean that system stability at the system scale, structure stability at the mesoscale and interaction between elements at the element scale can be distinguished and correlated according to the exact logic of real systems. Therefore, the EMMS paradigm allows the possibility of developing corresponding software and hardware that follow the same structure and logic as those of models and systems. We believe that such a paradigm of computation is one of optimized strategies to study complex systems in nature, engineering, and society. Because of the generality of multiscale structures and behaviors, this should be a direction of research in computational science and technology. Here, $x$PU represents the hardware defined by the EMMS paradigm. Currently, CPUs are suitable for computation at system scales featuring complicated logics but not so extensive computation, while GPUs suit that at element scales characterized by simple but massive and parallel computation. There are no optimal processing units for mesoscales yet, implying the direction to $x$PU.

Obviously, the core advantage of the EMMS paradigm is its definition of stability conditions at the macroscale and the mesoscale. Since stability conditions rule the interactions at these two scales and correlations between different scales, simplification and optimization of computations becomes possible, however, calling for great efforts to materialize it.

# Chapter 5
# The EMMS Principle in Virtual Process Engineering

## 5.1 Virtual Process Engineering

Virtual Process Engineering (VPE), combines computer simulations with real time, online data to provide engineers with the benefits of so-called virtual reality. VPE will allow engineers to simulate industrial processes with high accuracy and to, in real time, compare simulation outputs with experimental results online, and visualize relevant results dynamically and in 3D [12, 47]. The realization of VPE will require considerable progress in the accuracy of physical modeling, as well as improvements in the capabilities of computing hardware and software. When we applied the EMMS drag model to CFD simulation, we recognized the potential of using the EMMS paradigm to explore VPE. Following the logic of "first calculate the global distribution of a system, then evaluate local (meso-) structure and finally examine dynamics" [18, 43], the predictability and scalability of CFD could be improved substantially because of structural and logical consistency between problem, model, software, and hardware [13, 14, 16]. That is, the global distribution in a reactor can be calculated with the global stability constraints, and then the local stationary structure can be evaluated using the mesoscale stability conditions. Consequently, the dynamic evolution of a system can by modeled by CFD using the evaluated initial conditions (local stationary structures).

The main features of the EMMS paradigm were described in the original book [3]. It is essential that the structures (configurations) of the model, software, and hardware are similar to that of the problem

J. Li and W. Huang, *Towards Mesoscience*, SpringerBriefs in Applied Sciences and Technology, DOI: 10.1007/978-3-642-41790-0_5,

to be solved. The consistency of the logics with respect to computation (interaction) and storage capacities and formats at individual scales, and correlation (communication) between scales should be maintained between these four factors.

## 5.2 The Present and Future of Virtual Process Engineering

Currently, CPUs are suitable for computation at the system scale dominated by complicated system stability, whereas GPUs are used for the element scale featuring intensive interaction between elements. However, an optimized processing unit for the mesoscale has not been developed. If this kind of processing unit can be designed by following the logic and structure of the EMMS paradigm, the efficiency of computation will be improved greatly. We term such a processing unit $x$PU [18]. The design of $x$PU calls for combined effort from a variety of fields, particularly computer technology. Currently, an available device, Many Integrated Core (MIC), roughly meets the requirements of mesoscale computation. However, it is still far from ideal $x$PU as defined in Fig. 4.5.

As an alternative, the EMMS paradigm was implemented by integrating CPUs and GPUs; that is, CPU for the system scale, CPU + GPUs for the mesoscale and GPUs (with CPUs for control) for the microscale. This led to the construction of a 1.0-petaflop supercomputer [16, 18], which has been used widely in different fields such as biology [62], materials science [63], chemical engineering [46, 64], and metallurgy [44].

Following the strategy of the EMMS paradigm, we established a VPE laboratory to explore the virtual reality of process engineering [47]. A photograph of this laboratory is presented in Fig. 5.1. The aim of this platform was to achieve a virtual reality environment for process engineering through high-accuracy simulation of industrial processes in real time, online comparison between the simulated and experimental results, and 3D dynamic visualization of relevant results [47].

Currently, the speed of computation is still far from real-time simulation. We found that a one-minute simulation of the whole loop of a CFB system required one week of computation. We are currently

**Fig. 5.1** Version 1.0 of VPE, was developed at IPE to demonstrate the use of a VPE, platform for designing industrial processes. On the *right* is an experimental unit. A screen with the same height as that of the unit is shown in the *middle*. A computer controls both the experiment and simulation to allow online comparison. Reprinted from [3], Copyright 2013, with permission from Springer

attempting to speed up computation. However, this capacity is already useful for process design and optimization [3, 18, 48], which is possible only when the mesoscale modeling is used.

Traditionally, we used CPU clusters for parallel computation, although we now use CPU + GPU for hybrid parallel computation. We aim to make a breakthrough in understanding mesoscale phenomena, and achieve similarity between problem, modeling, software, and hardware, which will allow us to realize VPE. At that time, it will be possible to run a computer to perform a real-time simulation of an industrial process with sufficient capability to control and optimize it online, onsite, and in real time, as depicted schematically in Fig. 5.2.

In this scenario, with the specified operating conditions, material properties, boundary conditions, and reactor configurations, the

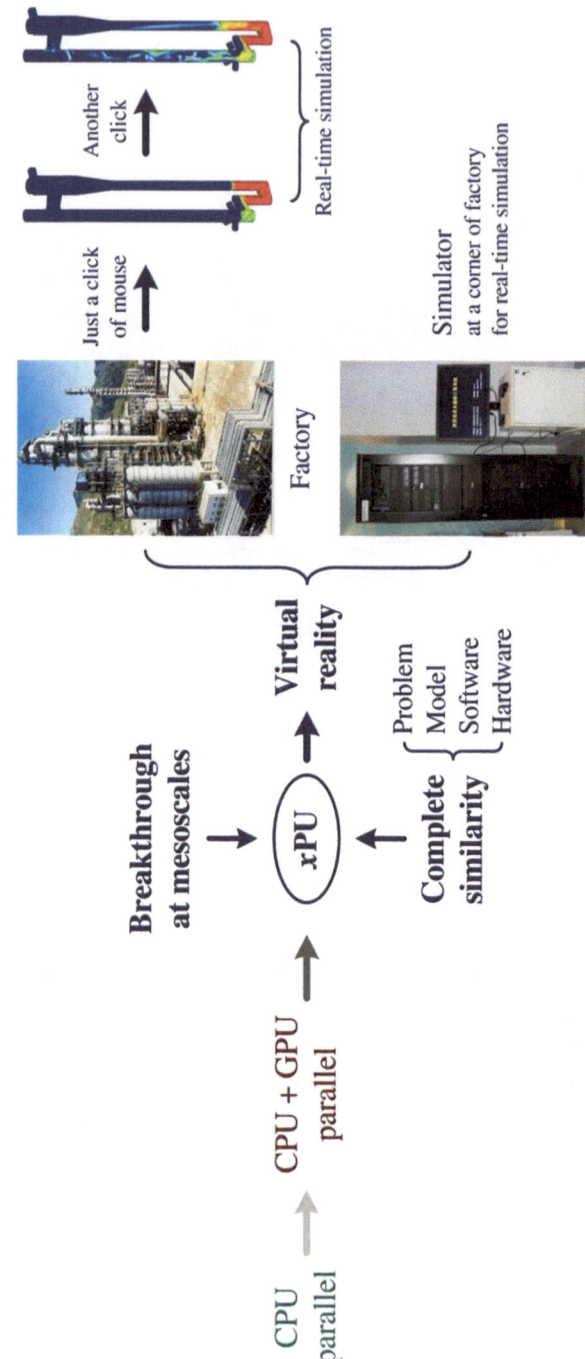

**Fig. 5.2** Computational power to realize virtual reality—past, present, and future

virtual reality inside a unit will be produced immediately after a click of a mouse, and the effect of a change in conditions may be visualized simply through another click. Whether or not this scenario can become a reality is totally subject to the progress of interdisciplinarity between computational science and process engineering. *x*PU and its integration with CPU and GPU will be the key.

# Chapter 6
# Towards Mesoscience

## 6.1 Current Limitations in Understanding Complex Systems

Since the publication of the EMMS model [5, 6] (better summarized in [19]) and its verification [7, 8], we have explored its extension and generality [15, 28, 29]. We confirmed its effectiveness in turbulence [11], and many other systems [12, 33, 37, 49, 65, 66], and clarified some relevant concepts by distinction of levels and scales [67]. Most importantly, the EMMS model was successfully used to solve many engineering problems, as reviewed in [3].

It occurred to us that the EMMS principle of compromise in competition might touch the limitations of current theories or even be what was missing between reductionism and holism. Because of the universality of the EMMS principle, and thus its importance, we believe that it is necessary to develop the interdisciplinary field of mesoscience, which could be based on the EMMS principle [3, 4, 50]. Although the relevant concepts are the same as those for mesoscale modeling, we have coined the term "mesoscience" to attract attention from different fields and promote it as a general interdisciplinary science, instead of limiting exploration to different specific mesoscale problems.

Recently, we investigated the principle of compromise in competition in more systems, including closing a turbulence model by considering the compromise in competition between viscous and inertial effects [49], protein folding to reveal the compromise in competition between minimum free energy and other mechanisms [65], and

J. Li and W. Huang, *Towards Mesoscience*, SpringerBriefs in Applied Sciences and Technology, DOI: 10.1007/978-3-642-41790-0_6,

material preparation to understand the compromise in competition between reaction and transport rates [66].

The EMMS principle could be applied to all of these diverse systems, which gave us confidence to propose the concept of mesoscience. Some may consider this too early with respect to our currently limited understanding, but we believe the best way to advance this field is to attract attention from the science and engineering communities by exploring a variety of mesoscale problems, which will be facilitated by clearly defining a term for this direction of research. In particular, we believe that when a common physical principle and corresponding unified mathematical formulation are identified, it is rational to explore the concept of a new interdisciplinary science, as discussed in [3, 4].

At the very least, triggering discussion on this topic will benefit our exploration of the best way to correlate element and system scales, even if mesoscience does not eventually become established.

## 6.2   The Mesoscales Between Element and System Scales

When we study a problem, we need to know what is the element, what is the system, and all external factors that influence the system through boundaries, as shown in Fig. 2.1. Figure 6.1 presents examples of systems relevant to chemical engineering: a bulk material consists of a huge number of atoms or molecules, a reactor consists of numerous particles, and a chemical complex consists of many individual reactors or units. The problems studied at these three levels could be on element, meso- or system scales. Gradually, we have accumulated much knowledge of both elements and systems, which include the traditional disciplines such as chemistry (materials), chemical engineering, and process system engineering. However, difficulties arose in the range between element and system scales because correlations between element scales and system scales, regardless of the level (material, reactor or system), are needed. We were also uncertain about how to integrate different disciplines (for instance, the three levels in Fig. 6.1) to optimize a whole chemical process from molecular to environmental scale. In fact, even if the problem studied is only on the element or system scale, it is still

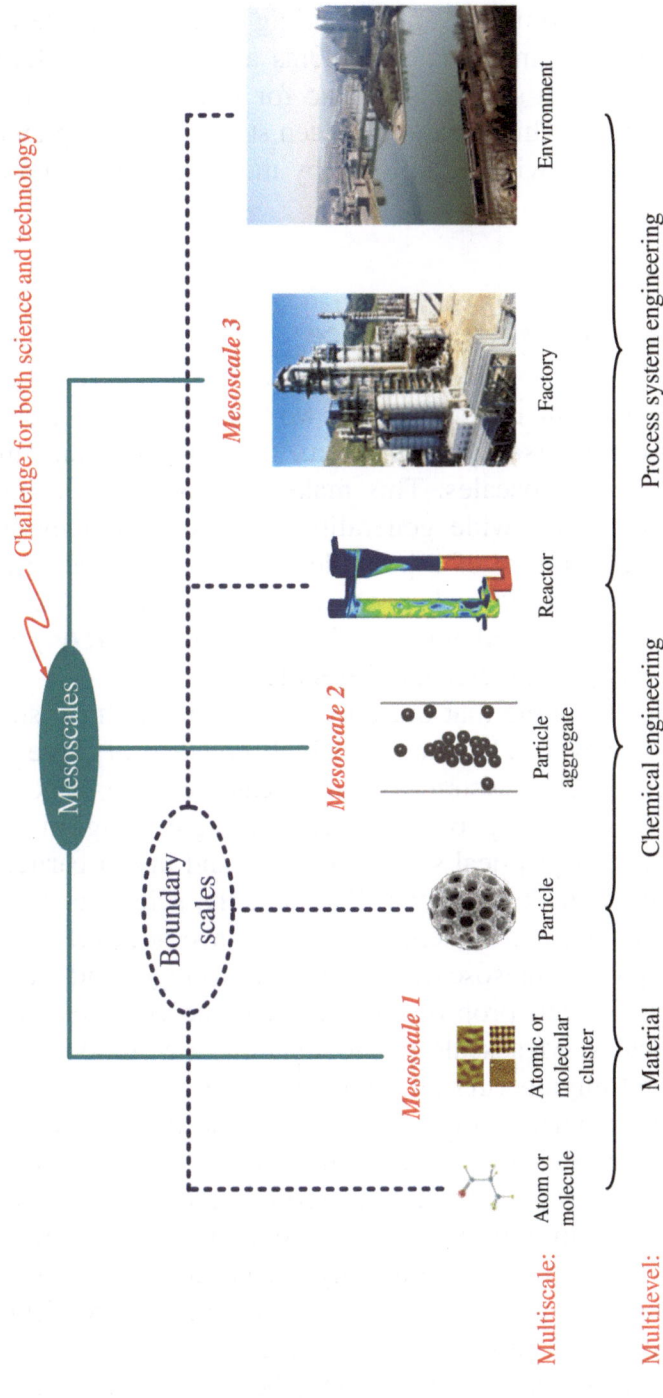

**Fig. 6.1** Three mesoscales at three levels commonly considered in chemical engineering. Reprinted from [3], Copyright 2013, with permission from Springer

difficult to have a complete understanding without considering the other scales. That is, mesoscale problems are a common challenge, not only for chemical science, but also for science and engineering as a whole [13, 14]. This is what has been studied by complexity science and we hope it will be explored by mesoscience from a different angle.

## 6.3  The Domain of Mesoscience

Based on our current understanding, we consider mesoscience the science of the universality of phenomena, processes, and mechanisms at various mesoscales. This makes mesoscience an interdisciplinary subject with wide generality. We proposed mesoscience based on the recognition of a possible common physical principle, *compromise in competition between dominant mechanisms*, and a unified framework of mathematical formulation, the *MOV problem*, for all mesoscale phenomena at different levels.

It is important to note that our universe is a hierarchical structure consisting of many levels, each of which consists of three scales: element (micro-) scale, system (macro-) scale, and the mesoscale in between, as shown in Fig. 6.1 for chemical engineering, in Fig. 6.3 for the spectrum of physical science [3, 50], and in our earlier work [7] for even broader descriptions. Traditionally, scientists have paid attention to element and system scales, but have neglected the governing principle at mesoscales, although average and statistical approaches have been proposed to describe the complex phenomena at mesoscales. Mesoscale phenomena are a common challenge for science and engineering as a whole [3, 48]. In fact, mesoscale issues exist everywhere, not only in nature but also in society. It is a common belief that the correlation between the microscale and macroscale is important, although the term "mesoscale" has not been widely used. This may be because the relationships between levels and scales in a system have not usually been specified so that several mesoscales might be blurred, leading to difficulties in identifying the governing principle on a specific mesoscale.

Most difficulties currently preventing the development of various individual disciplines and their integration occurs at various mesoscales;

however, they have not been tackled in a unified way. We have often been puzzled by the difficulty in formulating system behaviors from microscale mechanisms using approaches such as statistical mechanics, coarse-grained methods, and constitutive equations. In reality, we were often limited by mesoscale behaviors, but did not consider them from this viewpoint. That is, we did not understand that there might be a common principle and general theory that could describe different mesoscale problems. The importance of mesoscience is therefore its potential universality to explain all mesoscale phenomena existing between elemental particles and the observable universe.

One of the common characteristics of mesoscale phenomena is dynamic heterogeneity. With this common characteristic in mind, it is important to first identify the dominant mechanism promoting the formation of heterogeneity and that for maintaining uniformity. There could be either one or multiple mechanisms for each, but might be possible to integrate them into two representative mechanisms that govern these two respective effects.

Traditional sciences generally consider the boundary scales of elements and systems. Attempts to describe mesoscales have followed either reductionist or holistic approaches. In contrast, mesoscience will focus on the mesoscales that bridge boundary scales and use a combination of reductionism and holism to reveal a universal principle governing mesoscales.

What should be noted is the relationship between *ab initio* approaches and mesoscience. If we understand all of the information at element scales, and have supercomputing capability to perform computations based on element models, we will be able to reproduce system behavior including mesoscale phenomena. In this case, the compromise in competition would also be generated. However, such a task requires vast computational power. The role of mesoscience is to simplify such problems using a generalized principle, which should allow them to be simulated accurately at reduced computational cost. That is, recognition of the governing principle in the formation of mesoscale phenomena will greatly simplify the analysis of them. This is the importance of mesoscience.

We would expect that a generalized governing principle for mesoscales will be used to upgrade the traditional methods applied in different coarse-grained and statistical models in various fields.

## 6.4  A Common Governing Rule of Mesoscience

Our research indicates that a common principle may exist for all
mesoscale problems, that is, "compromise in competition between
dominant mechanisms with coordination" in physics and the "MOV
formulation" in mathematics [3, 4, 50]. Such universalities in both
physics and mathematics have been analyzed for the three levels of
chemical engineering (Fig. 6.1) and can potentially be extended to
all mesoscales in different fields, which promoted us to explore the
concept of mesoscience, as detailed recently [3]. Mesoscience may
act as a bridge between reductionism to understand information at
"elemental" or "small" scales and holism to interpret global behavior at "system" or "large" scales; that is, it is a science between
small and large scales, or between holism and reductionism. Here,
"small" and "large" are both relative concepts, spanning the range
of size between elemental particles and the universe, giving mesoscience an interdisciplinary nature. In fact, social and economic
sciences have also paid much attention to the importance of compromise, competition, and coordination [68, 69]. We believe that
mesoscience may allow us to unify the natural and social sciences
because the common principles of these two branches of science
occur at mesoscales, even though the elemental and system scales
involved are totally different.

Clarifying the mesoscales considered in each discipline will
hopefully allow us to understand the domain of each discipline and
the relationships between them, which should promote the interdisciplinarity between traditional sciences. For example, chemistry and chemical engineering have the particle scale as a common
boundary scale, as shown in Fig. 6.1; it is the system scale for
chemistry but the element scale for chemical engineering. Only
when the mesoscales of chemistry and chemical engineering are
understood can these two disciplines be integrated via the particle
scale. An added complication is that currently chemists and chemical engineers understand the particle scale differently and even use
different terminology and parameters. Chemists pay main attention
to the parameters of reactions, while chemical engineers consider
transport parameters much more.

## 6.5  The Transdisciplinary Nature of Mesoscience

Mesoscience is not a new topic. The involved problems and mechanisms are the same, but the angle from which they are viewed is different from traditional approaches when considering the concept of mesoscales. Changing the approach used to solve old problems can lead to progress in solving them, but calls for interdisciplinary collaboration. That is, traditional knowledge and approaches need to be integrated with those from different disciplines, and in particular, the principle of compromise in competition should be included. All fields of research may contribute respective specific knowledge at corresponding levels to reveal the common principles that underpin mesoscience [3]. Although different disciplines consider their unique problems, the compromise in competition with coordination will be their common language, and MOV problems will be a common mathematical tool. That is, concrete dominant mechanisms are discipline-specific, but their physical principle and mathematical formulation for compromise in competition will be identical. Mesoscience will strongly promote interdisciplinarity. This universality is the rationale for mesoscience, without which it is not necessary to have a new concept.

## 6.6  The Levels and Scales in Systems

It is critical to recognize that mesoscale issues can be correctly analyzed only when "levels" are properly identified. Without correct identification of "levels," the compromises in competition at different levels may be blurred, making it impossible to extract the level-specific dominant mechanisms and their compromise [67], as shown in Fig. 6.2. That is, a general correlation between microscale and macroscale without specifying their corresponding level (that is, defining the element and system scales) is neither scientific nor sufficient. Neglect of this important point may be the reason why the universality of the governing rules for all mesoscales was not recognized earlier. In fact, either blurring two or more mesoscales or missing part of

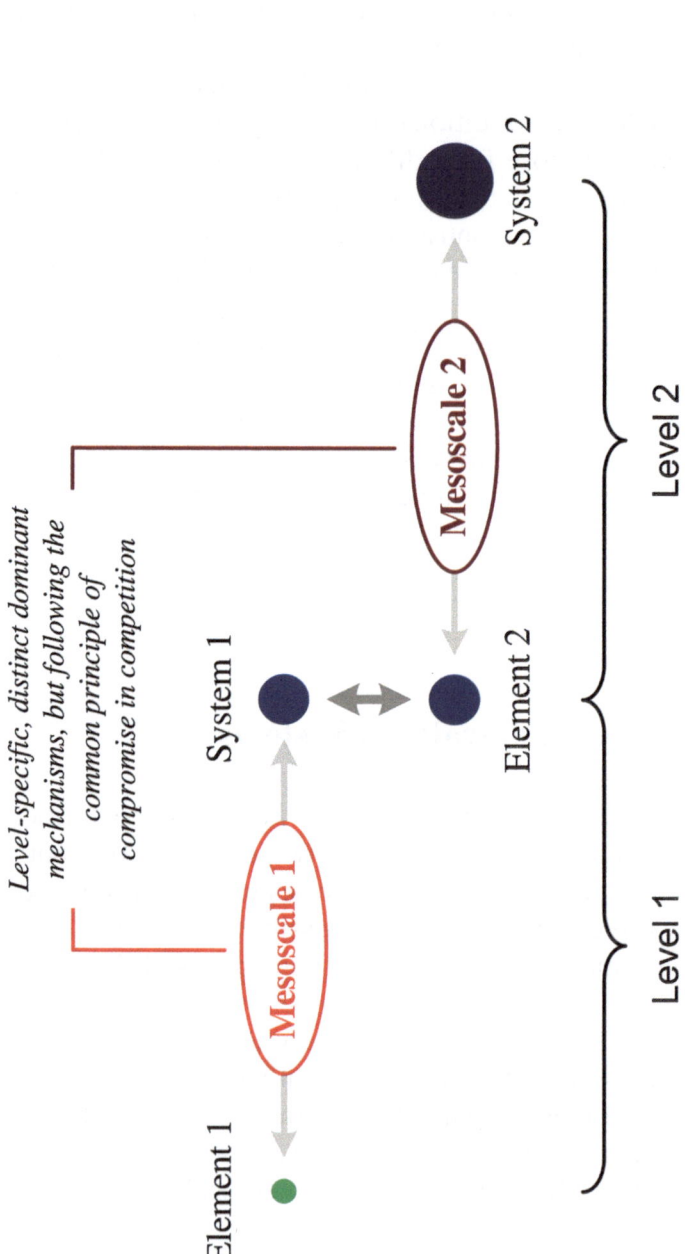

**Fig. 6.2** Illustration of various mesoscales at hierarchical levels

a single mesoscale during analysis makes it impossible to identify the specific dominant mechanisms, (cf. Fig. 6.2). Each dominant mechanism in compromise always corresponds to a specific mesoscale that is influenced by both elemental and system scales, but not directly by the neighboring level. Therefore, without correct identification of the related level, the principle of compromise in competition cannot be applied. At least, distinguishing levels makes the analysis of complex systems much simpler.

Therefore, identifying levels should be the first step when considering mesoscale issues. Distinction of levels and scales is also important in traditional disciplines, because any individual boundary scale is in the domains of two disciplines, one as the element scale, and the other as the system. Complete understanding of this scale therefore requires integration of knowledge from both sides, to which sufficient attention has not been paid in most cases, for example, the particle scale for chemists and chemical engineers. Therefore, in addition to understanding mesoscale issues, integration of the knowledge from different disciplines at boundary scales is also important.

The description of levels and scales in Fig. 6.3 is only preliminary. For instance, the level between an atom and bulk material particles should be divided into two levels if atomic clusters form nanoscale units first, and then these units assemble themselves into bulk material particles [70]. The principle of compromise in competition should be still valid for both levels even though different dominant mechanisms could be involved. Additionally, it is also possible to have different compromise processes at a single level, for example, corresponding to the structures and properties of materials. We know that it is important to correlate microscale phenomena with macroscale behaviors, but usually fail to emphasize how many levels exist between these microscales and macroscales, and which level we are working on. The correlation between levels can be realized only when the mesoscales at these levels are all resolved.

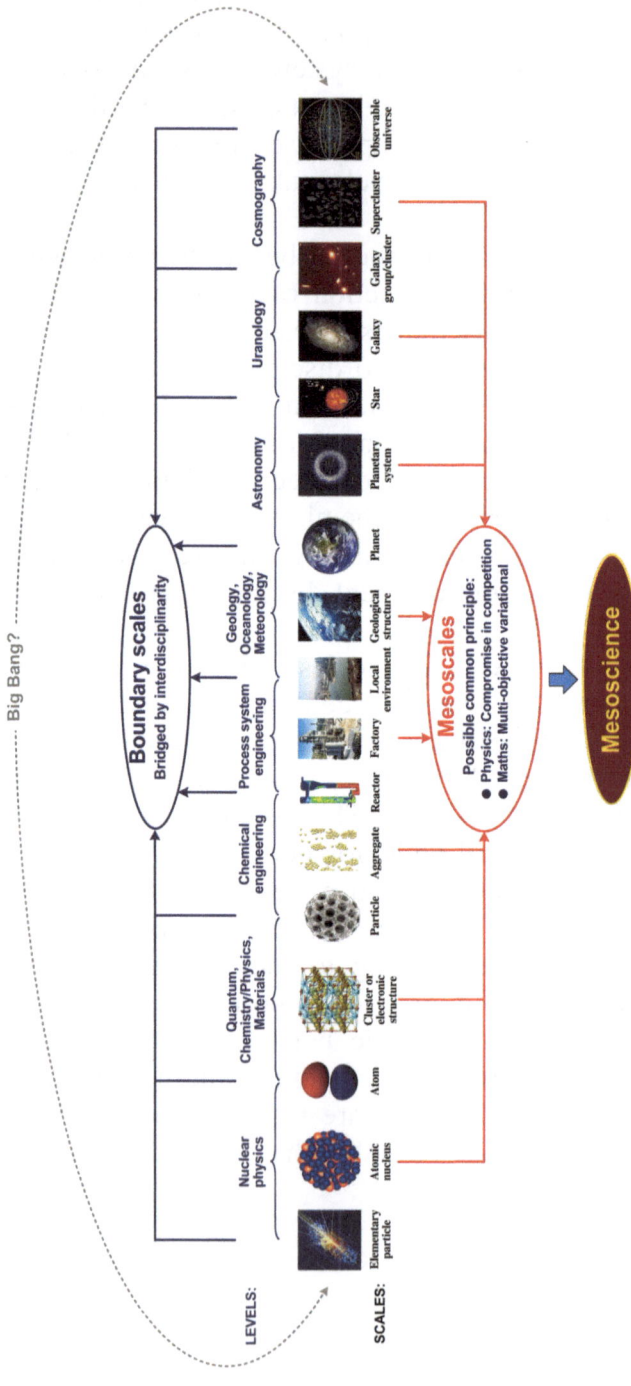

**Fig. 6.3** A unified theory of mesoscience encompasses all mesoscale phenomena. Modified from [3], Copyright 2013, with permission from Springer

## 6.7  A Preliminary Definition of Mesoscience

Figure 6.3 is a preliminary summary of the concept of mesoscience, showing some important features of an interdisciplinary science; that is, common principles and physical and mathematical arguments. From these, preliminary definitions of mesoscience could be "the science of the universality of mesoscale phenomena at different levels," or "the science of compromise in competition with coordination," or "the science of compromise in competition," or simply the science of "the in-between [4]," which all involve identifying common physical principles and their mathematical formulation [3, 4] to address different problems. Solution of a problem using mesoscience should start with identification of a level. The element and system should then be defined, and the dominant mechanisms and their compromise are analyzed. Finally, the phenomena should be formulated and solved with a MOV problem, as preliminarily summarized as well in Sect. 4.3. Currently, simplification is needed in dealing with MOV problems.

# Chapter 7
# Perspectives of Mesoscience

Scientists in different fields know the discipline-specific challenges they face. However, if they are asked what the most challenging problem in common for scientists and engineers is, their answers are diverse. Chemists might mention molecular self-assembly, physicists would think of the correlative electron problem, and biologists could say protein structures. If they are further asked for a word that reflects the common nature of all of these challenges, their answers would also be diverse. However, if someone proposed heterogeneity (in space and in time) as a suitable word, they would readily accept that it is really a common challenge for different fields because we live in a heterogeneous world.

The above discussion shows that different disciplines share a common nature—heterogeneity. This is certainly the driving force for complexity science focusing on the correlation between microscale and macroscale, which is complicated by heterogeneity. When faced with heterogeneity, researchers usually either use simplified solutions with averaging assumptions or try to know all the details, overlooking the most important dominant mechanisms, and thus unable to describe complex systems accurately with sufficient scalability. In fact, heterogeneity has been challenging scientists and engineers in almost all fields, but is usually treated with averaging approaches. For example, "wave functions" are employed to describe the statistical behavior of quantum particles, density functional theory uses a fictitious single particle to approximate many-body effects [71, 72], and coarse-graining is adopted in cosmology [73]. From the viewpoint of mesoscience, averaging approaches have neglected the governing principle for heterogeneity at the mesoscales.

J. Li and W. Huang, *Towards Mesoscience*, SpringerBriefs in Applied Sciences and Technology, DOI: 10.1007/978-3-642-41790-0_7,

This naturally led us to ask: is there a common governing principle that underlies the diverse heterogeneities present in different systems, and what is it? This is a major challenge in contemporary science and technology because we understand that different heterogeneities are based on different dominant mechanisms. In this case, direct study of the common nature of different systems is difficult, and to further look for a unifying principle is even harder. Indirect study of individual systems paying attention to extracting common principles would be a simpler approach.

Our study has followed the strategy from individual to general. From this work, we recognized that a common principle was missing at all mesoscales in previous studies of heterogeneities. That is, heterogeneities are related to mesoscale phenomena, and subject to the principle of compromise in competition with coordination.

It is interesting that the traditional, popular term "heterogeneity" is closely related to the new "mesoscience" through the bridge of the principle of "compromise in competition." Therefore, mesoscience is the science to explain variations of heterogeneity with the principle of compromise in competition, because heterogeneity always occurs at mesoscales.

A variety of concepts related to heterogeneity in time or space are summarized in Fig. 7.1. In natural or artificial systems, we encounter various entities of complexity or diversity. Relevant phenomena appear frequently such as many-body effects, collective behavior and chaos, and the features of these phenomena might be nonlinear, nonequilibrium, critical (bifurcation or transition) and so on. Previous theoretical efforts to tackle such problems include renormalization [74], dissipative structures [75], and synergetics [76], while methodological efforts involve coarse graining, statistics and many others. The common features of such complexity are spatial or temporal heterogeneities, which emerge at mesoscales because of the compromise between competing mechanisms with coordination, as expressed in MOV formulations. Compared with simple or linear systems, complexity or nonlinearity invokes the compromise between multiple simple processes. Therefore, the EMMS principle of compromise in competition, or something equivalent, should be used in addition to traditional theories to solve problems. In short, heterogeneity is a common challenge, and understanding it is the objective of

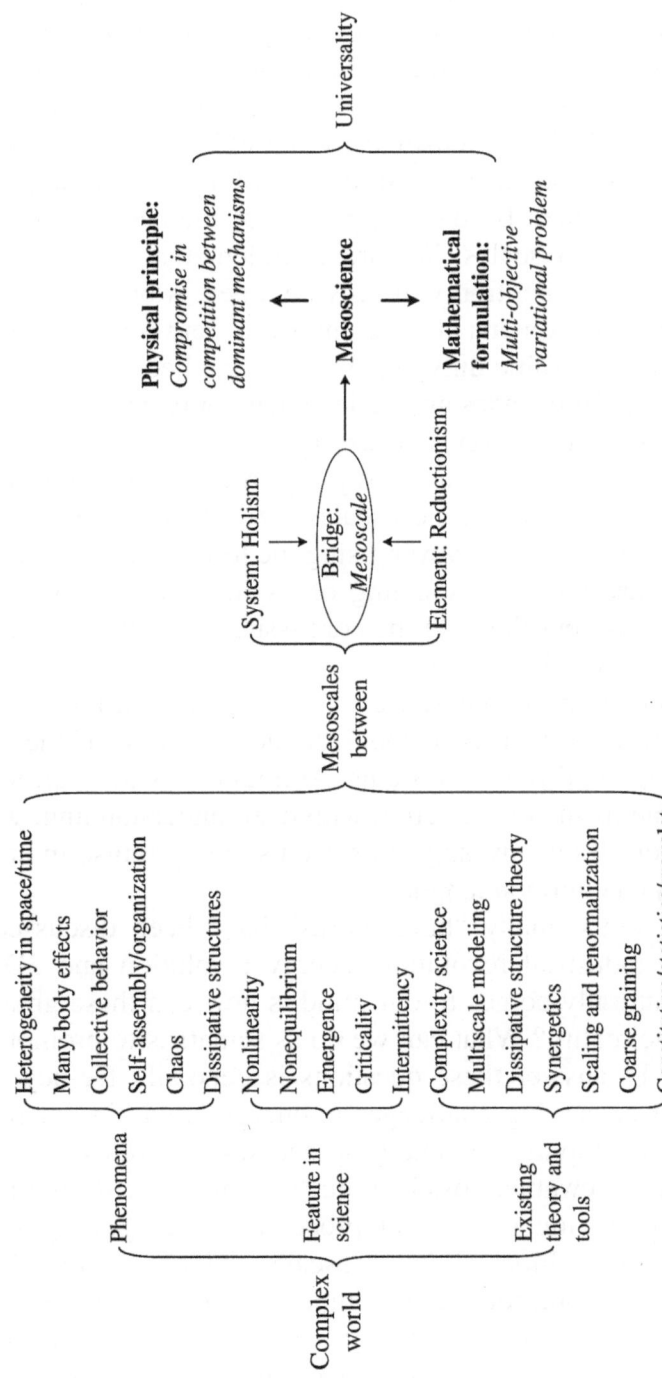

**Fig. 7.1** Summary of the complexity at mesoscales

mesoscience. Therefore, mesoscience is not a new concept, but unification of simplicity and complexity and integration of holism and reductionism through the principle of compromise in competition.

Our confidence in the concept of mesoscience was established by the predictability and scalability of the EMMS model in CFD computations [26, 33, 37], intensified by the effectiveness of the principle of compromise in competition in solving different problems [11, 12], and clarified by the recognition of the need to distinguish between levels and scales. We are confident that the principle of compromise in competition will provide a new way to analyze various mesoscale phenomena at different levels between elemental particles and the observable universe. We have studied problems related to mesoscience for decades and believe that applying the principle of compromise in competition is necessary.

In essence, there is only a small jump from considering the common principle for different mesoscales [28, 29] to using the term of mesoscience [3, 4, 50]. However, using the term of mesoscience may help to increase the understanding of its importance from specific problems in different fields to the core issues that affect science and technology as a whole.

Our participation in conferences, and reviewing various proposals and papers has allowed us to look for the relevance of the EMMS principle in many different systems. Although it is difficult to relate it to all problems in consideration with deep understanding, we have found that for all our investigated systems, compromise in competition is always a common language.

In recent years, many "buzz words" have been discussed such as the third industrial revolution, energy revolution and 3D printing. What naturally comes to our mind is how can these grand concepts become reality? What can we do as scientists to promote these transitions? To answer these questions is generally far beyond our capacity, but discussing challenges in energy is within our research domain. In developing new energy processes, regardless of whether they involve renewable, fossil or nuclear energies, designing new materials, optimizing reactors or processes or integrating systems, progress is always blocked at mesoscales, which are all subject to the principle of compromise in competition and MOV formulation [18]. Therefore, whether an energy revolution can be realized will probably depend on the advances made in understanding mesoscale phenomena at the three levels of chemical science: material, reactor, and system, as shown in Fig. 6.1. That is, we believe mesoscience will contribute much to the anticipated changes in the field of energy.

Recently, we used the EMMS principle of compromise in competition to analyze the stability condition of protein folding [65, 77] and in a turbulence model [49]. Protein folding is a dynamic process in which the free energy minimum dominates; however, is not exclusive, but compromises with other competing factors. It is this compromise that keeps the protein dynamic, and hence, active. We recognized that the analysis of the free energy landscape alone is not sufficient, and the time evolution of free energy needs to be taken into account [77]. Although this is to be confirmed further in detail, the critical importance of compromise in competition between dominant mechanisms deserves great attention.

Since the seminal work [78] of Turing, reaction-diffusion models have been deployed extensively to describe self-organization phenomena. In such systems, it can be roughly argued that spatially or temporally heterogeneous structures form only when reaction processes compromise with transfer processes. Figure 7.2 illustrates an example of the Gray-Scott model [79]. Phenomenologically, it is interesting to observe that the complex structures appear within the parameter range where the mass transfer rate is comparable with the reaction rate. If we can establish a stability condition for such systems from the principle of compromise in competition, we may manipulate the patterns easily.

We believe that correlated electron systems are also subject to the principle of compromise in competition. Decreasing particle size to the nanoscale (for instance, gold nanoparticles [80]), or decreasing a system's temperature, changes the compromise in competition within materials, which leads to a change in collective behaviors such as conductivity.

Mesoscales have been investigated in fields other than chemical engineering, as reviewed in [3, 4, 50]. The term "mesoscale" has been widely adopted in meteorology [81–83]. For example, Mesoscale Meteorological Models are well known, and "mesoscale meteorology" appears in Wikipedia [84]. In chemistry and materials science, "mesoscale" is equivalent to "mesoscopic scale," which refers to a specific spatial range. For example, mesoscale has been used to describe the ranges of 2–50 nm [85], 10–100 nm [86], 1–1,000 nm [87], or "approximately an order of magnitude larger than the nanoscale"

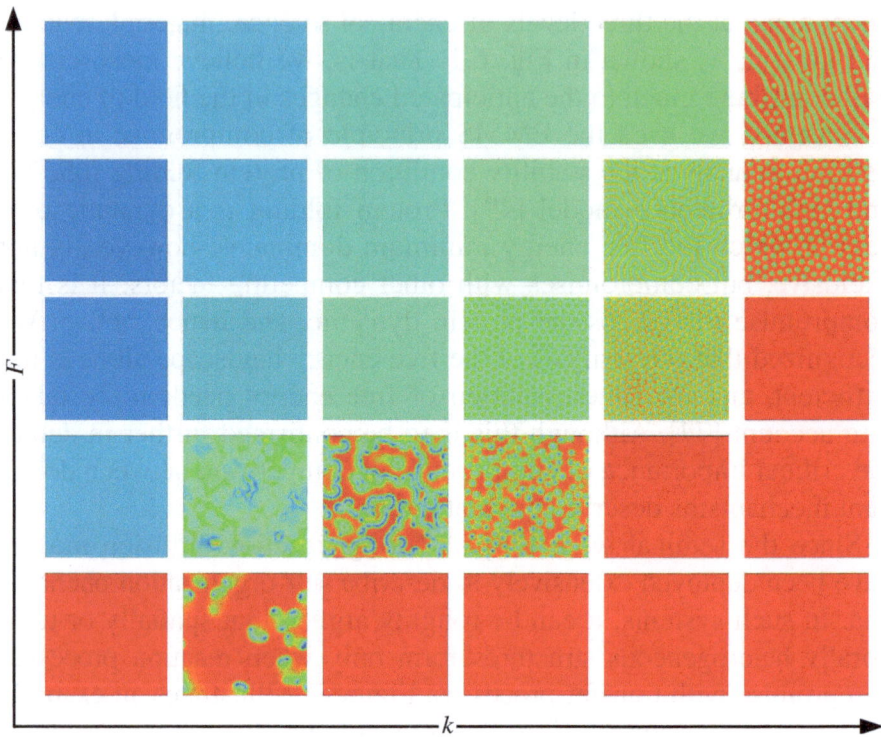

**Fig. 7.2** Patterns of the Gray-Scott model within the parameter plane of reaction rate constant (*k*) and mass transfer rate (*F*). Modified from [3], Copyright 2013, with permission from Springer

[88]. Schmalian and Wolynes [89] recognized the mesoscale analogy between strongly correlated electronic states and colloidal systems. In [70], however, "meso" was related to an "in-between" principle of operation. Mesoscale science has been proposed in chemical biology with systems approaches [90]. Some terms related to mesoscale are mesostructure [87, 91], mesocrystal [92], meso compound [93], meso level [94] and mesophase [86, 95]. The term mesoscience used in [94] means something different from what we describe here. It is also worth noting that different fields use different nomenclature to describe mesoscale phenomena. For example, in biology investigations involving population dynamics [96], the "averaging problem" in cosmology [73], and self-assembly at various scales with similar, but not identical, rules [97], all, in fact, investigate mesoscale phenomena. The terms "self-organization" or "emergence" are usually used to

describe the phenomena or processes [98, 99], while theoretical efforts refer to dissipative structures [75] and synergetics [76]. In addition, mesoscience has been used in journal and organization names, indicating the increasing attention to this topic from different fields.

It is far beyond our capacity to analyze the compromise process in different systems that are not related to chemical engineering. However, it is reasonable to question why this principle has been overlooked in the current study of many fields.

We are glad to see that increasing attention is being paid to mesoscience from important international foundations. For example, the National Natural Science Foundation of China (NSFC) has recently launched a mesoscience program focusing on the two mesoscales of chemical engineering at the material level and the reactor level [100], that is, the mesoscale between atom or molecule and bulk marerials and that between single particle and reactor. The U.S. Department of Energy (DOE) is planning a program on mesoscale science at the material level [101], with interest on superconductivity, magnetism, rigidity, melting, energy-related devices, materials failure, assembling and so on within the mesoscopic range [2] of physical size corresponding to mesoscopic physics.

Focusing on various mesoscales in science and engineering, particularly, with special attention to the corresponding governing principles for them will enable the progress in multiscale modeling and computation.

In response to this increased understanding, we proposed the concept of mesoscience to emphasize its universality and importance in a book entitled "From multiscale modeling to mesoscience—a chemical engineering perspective." This subsequent brief for the book cleans up our understanding of this concept. We hope to make a step from mesoscales to mesoscience, and welcome all comments from colleagues to engender discussion and improve our understanding of this concept.

When we began our research we never expected to find that the traditional terms "heterogeneity" and "meso" are correlated. We also never thought that this correlation would become a reality with the EMMS principle acting as a bridge. All of these findings were adventitious, but we now believe that the concept of compromise in competition will enable further progress to be made in science and technology.

# References

1. http://en.wikipedia.org/wiki/Complexity_science
2. http://en.wikipedia.org/wiki/Mesoscopic_physics
3. Li J, Ge W, Wang W, Yang N, Liu X, Wang L, He X, Wang X, Wang J, Kwauk M (2013) From multiscale modeling to meso–science—a chemical engineering perspective. Springer, Berlin
4. Li J, Huang W, Edwards PP, Kwauk M, Houghton JT, Slocombe D (2013) On the universality of mesoscience: science of 'the in–between'. arXiv:1302.5861v1
5. Li J, Tung Y, Kwauk M (1988) Method of energy minimization in multi–scale modeling of particle–fluid two–phase flow. In: Basu P, Large JF (eds) Circulating fluidized bed technology II. Pergamon, London, pp 89–103
6. Li J (1987) Energy–minimized multi–scale model for particle–fluid two–phase flow. Ph. D. Dissertation, Institute of Chemical Metallurgy, Chinese Academy of Sciences, Beijing, China
7. Li J, Zhang J, Ge W, Liu X (2004) Multi–scale methodology for complex systems. Chem Eng Sci 59:1687–1700
8. Zhang J, Ge W, Li J (2005) Simulation of heterogeneous structures and analysis of energy consumption in particle–fluid systems with pseudo–particle modeling. Chem Eng Sci 60:3091–3099
9. Ge W, Li J (1997) Pseudo–particle approach to hydrodynamics of particle–fluid systems. In: Kwauk M, Li J (eds) Circulating fluidized bed technology V. Science Press, Beijing, pp 260–265
10. Ge W, Li J (2003) Macro–scale phenomena reproduced in microscopic systems—pseudo–particle modeling of fluidization. Chem Eng Sci 58:1565–1585
11. Li J, Zhang Z, Ge W, Sun Q, Yuan J (1999) A simple variational criterion for turbulent flow in pipe. Chem Eng Sci 54:1151–1154
12. Ge W, Chen F, Gao J, Gao S, Huang J, Liu X, Ren Y, Sun Q, Wang L, Wang W, Yang N, Zhang J, Zhao H, Zhou G, Li J (2007) Analytical multi–scale method for multi–phase complex systems in process engineering—bridging reductionism and holism. Chem Eng Sci 62:3346–3377
13. Li J, Ge W, Kwauk M (2009) Meso-scale phenomena from compromise—a common challenge, not only for chemical engineering. arXiv:0912.5407v3

J. Li and W. Huang, *Towards Mesoscience*, SpringerBriefs in Applied Sciences and Technology, DOI: 10.1007/978-3-642-41790-0,
© The Author(s) 2014

14. Li J, Ge W, Wang W, Yang N (2010) Focusing on the meso–scales of multi–scale phenomena—in search for a new paradigm in chemical engineering. Particuology 8:634–639

15. Li J, Kwauk M (2003) Exploring complex systems in chemical engineering—the multi–scale methodology. Chem Eng Sci 58:521–535

16. Chen F, Ge W, Guo L, He X, Li B, Li J, Li X, Wang X, Yuan X (2009) Multi–scale HPC system for multi–scale discrete simulation—development and application of a supercomputer with 1 Petaflops peak performance in single precision. Particuology 7:332–335

17. http://www.top500.org/system/176899

18. Ge W, Wang W, Yang N, Li J, Kwauk M, Chen F, Chen J, Fang X, Guo L, He X, Liu X, Liu Y, Lu B, Wang J, Wang J, Wang L, Wang X, Xiong Q, Xu M, Deng L, Han Y, Hou C, Hua L, Huang W, Li B, Li C, Li F, Ren Y, Xu J, Zhang N, Zhang Y, Zhou G, Zhou G (2011) Meso–scale oriented simulation towards virtual process engineering (VPE)—the EMMS paradigm. Chem Eng Sci 66:4426–4458

19. Li J, Kwauk M (1994) Particle–fluid two–phase flow: the energy–minimization multiscale method. Metallurgical Industry Press, Beijing (Available at http://www.emms.cn/res_base/emms_en_mpcs_cn/upload/channel/file/2013_3/8_5/qq2yhjzhto0x.pdf)

20. Li J, Reh L, Kwauk M (1990) Application of energy minimization principle to hydrodynamics of circulating fluidized beds. In: Basu P, Horio M, Hasatani M (eds) Circulating fluidized bed technology III. Pergamon, London, pp 105–111

21. Li J, Kwauk M, Reh L (1992) Role of energy minimization in gas/solid fluidization. In: Potter OE, Nicklin DJ (eds) Fluidization VII. Engineering Foundation, New York, pp 83–91

22. Li J, Chen A, Yan Z, Xu G, Zhang X (1993) Particle–fluid contacting in dense and dilute phases of circulating fluidized beds. In: Avidan AA (ed) Circulating fluidized bed technology IV. Pergamon, New York, pp 48–53

23. Li J, Wen L, Ge W, Cui H, Ren J (1998) Dissipative structure in concurrent–up gas–solid flow. Chem Eng Sci 53:3367–3379

24. Qian G, Sun G, Wen L, Li J (1998) Micro–scale measurement of dynamic flow structure in circulating fluidized beds. Powder Technol 100:76–77

25. Li J, Zhang X, Zhu J, Li J (1998) Effects of clusters behavior on gas–solid mass transfer in CFB. In: Fan LS, Knowlton T (eds) Fluidization IX. Engineering Foundation, New York, p 405

26. Li J, Cheng C, Zhang Z, Yuan J, Nemet A, Fett FN (1999) The EMMS model—its application, development and updated concepts. Chem Eng Sci 54:5409–5425

27. Cui H, Li J, Kwauk M, An H, Chen M, Ma Z, Wu G (2000) Dynamic behaviors of heterogeneous flow structure in gas–solid fluidization. Powder Technol 112:7–23

28. Li J (2000) Compromise and resolution—exploring the multi–scale nature of gas–solid fluidization. Powder Technol 111:50–59

29. Li J, Kwauk M (2001) Multi–scale nature of complex fluid–particle system. Ind Eng Chem Res 40:4227–4237
30. Liu M, Li J, Kwauk M (2001) Application of the energy–minimization multi–scale method to gas–liquid–solid fluidized beds. Chem Eng Sci 56:6805–6812
31. Ren J, Mao Q, Li J, Lin W (2001) Wavelet analysis of dynamic behavior in fluidized beds. Chem Eng Sci 56:981–988
32. Ge W, Li J (2002) Physical mapping of fluidization regimes—the EMMS approach. Chem Eng Sci 57:3993–4004
33. Yang N, Wang W, Ge W, Li J (2003) CFD simulation of concurrent–up gas–solid flow in circulating fluidized beds with structure–dependent drag coefficient. Chem Eng J 96:71–80
34. Yang N, Wang W, Ge W, Li J (2003) Choosing structure–dependent drag coefficient in modeling gas–solid two–phase flow. China Particuology 1:38–41
35. Wang L, Jin D, Li J (2003) Effect of dynamic change of flow structure on mass transfer between gas and particles. Chem Eng Sci 58:5373–5377
36. Ma J, Ge W, Wang X, Wang J, Li J (2006) High–resolution simulation of gas–solid suspension using macro–scale particle methods. Chem Eng Sci 61:7096–7106
37. Wang W, Li J (2007) Simulation of gas–solid two–phase flow by a multi–scale CFD approach—extension of the EMMS model to the sub–grid level. Chem Eng Sci 62:208–231
38. Yang N, Chen J, Zhao H, Ge W, Li J (2007) Explorations on the multi–scale flow structure and stability condition in bubble columns. Chem Eng Sci 62:6978–6991
39. Lu B, Wang W, Li J, Wang X, Gao S, Lu W, Xu Y, Long J (2007) Multi–scale CFD simulation of gas–solid flow in MIP reactors with a structure–dependent drag model. Chem Eng Sci 62:5487–5494
40. Dong W, Wang W, Li J (2008) A multiscale mass transfer model for gas–solid riser flows: Part 1—Sub–grid model and simple tests. Chem Eng Sci 63:2798–2810
41. Dong W, Wang W, Li J (2008) A multiscale mass transfer model for gas–solid riser flows: Part II—Sub–grid simulation of ozone decomposition. Chem Eng Sci 63:2811–2823
42. Zhang N, Lu B, Wang W, Li J (2008) Virtual experimentation through 3D full–loop simulation of a circulating fluidized bed. Particuology 6:529–539
43. Liu Y, Chen J, Ge W, Wang J, Wang W (2011) Acceleration of CFD simulation of gas–solid flow by coupling macro–/meso–scale EMMS model. Powder Technol 212:289–295
44. Xu J, Qi H, Fang X, Lu L, Ge W, Wang X, Xu M, Chen F, He X, Li J (2011) Quasi–real–time simulation of rotating drum using discrete element method with parallel GPU computing. Particuology 9:446–450
45. Yang N, Wu Z, Chen J, Wang Y, Li J (2011) Multi–scale analysis of gas–liquid interaction and CFD simulation of gas–liquid flow in bubble columns. Chem Eng Sci 66:3212–3222

46. Xiong Q, Li B, Zhou G, Fang X, Xu J, Wang J, He X, Wang X, Wang L, Ge W, Li J (2012) Large–scale DNS of gas–solid flows on Mole–8.5. Chem Eng Sci 71(26):422–430

47. Liu X, Guo L, Xia Z, Lu B, Zhao M, Meng F, Li Z, Li J (2012) Harnessing the power of virtual reality. Chem Eng Prog 108(7):28–33

48. Wang W, Ge W, Yang N, Li J (2012) Meso: the next big thing in simulation? Chem Eng 852(6):40–43

49. Zhang L, Qiu X, Wang L, Li J (2013) A stability condition for the turbulence model: from EMMS model to EMMS–based turbulence model. arXiv:1311.1272v1

50. Li J (2013) From customized multiscale modeling to general mesosicence—the principle of compromise. In: Yu A, Dong K, Yang R, Luding S (eds) Powders and grains 2013. AIP Conf Proc, vol 1542, pp 7–11

51. http://en.wikipedia.org/wiki/Standard_Model

52. http://en.wikipedia.org/wiki/Galaxy

53. http://en.wikipedia.org/wiki/Supercluster

54. Huang WL, Ge W, Li C, Hou C, Wang X, He X (2012) Atomic and electronic structures of Si[0 0 1](1 3 0) symmetric tilt grain boundaries based on first–principles calculations. Comput Mater Sci 58:38–44

55. Gale PA (2007) Supramolecular chemistry anniversary. Chem Soc Rev 36:141–142

56. http://en.wikipedia.org/wiki/Geology

57. http://en.wikipedia.org/wiki/Astronomy

58. http://en.wikipedia.org/wiki/Biology

59. Horgan J (1995) From complexity to perplexity. Sci Am 272:74–79

60. Li J, Wen L, Qian G, Cui H, Kwauk M, Schouten JC, Van den Bleek CM (1996) Structure heterogeneity, regime multiplicity and nonlinear behavior in particle–fluid systems. Chem Eng Sci 51:2693–2698

61. Xu G, Li J (1998) Analytical solution of the energy–minimization multi–scale model for gas–solid two–phase flow. Chem Eng Sci 53:1349–1366

62. Xu J, Wang X, He X, Ren Y, Ge W, Li J (2011) Application of the Mole–8.5 supercomputer: probing the whole influenza virion at the atomic level. Chin Sci Bull 56:2114–2118

63. Hou C, Xu J, Wang P, Huang W, Wang X, Ge W, He X, Guo L, Li J (2013) Petascale molecular dynamics simulation of crystalline silicon on Tianhe–1A. Int J High Perform Comput Appl. doi:10.1177/1094342012456047

64. Chen F, Ge W, Li J (2009) Molecular dynamics simulation of complex multiphase flow on a computer cluster with GPUs. Sci China–Chem 52(3):372–380

65. Xu J, Ren Y, Li J (2013) Multi–scale simulations of protein folding: application to formation of secondary structures. J Biomol Struct Dyn. doi:10.1080/07391102.2012.709461

66. Wang H, Han Y, Li J (2013) Dominant role of compromise between diffusion and reaction in the formation of snow–shaped vaterite. Cryst Growth Des 13:1820–1825

67. Li J, Ge W, Zhang J, Kwauk M (2005) Multi–scale compromise and multi–level correlation in complex systems. Chem Eng Res Des 83(A6):574–582
68. Pluchino A, Latora V, Rapisarda A (2006) Compromise and synchronization in opinion dynamics. arXiv:physics/0510141
69. Ben–Naim E, Krapivsky PL, Redner S (2002) Bifurcations and patterns in compromise processes. arXiv:cond–mat/0212313v1
70. Antonietti M, Ozin GA (2004) Promises and problems of mesoscale materials chemistry or why meso? Chem Eur J 10:28–41
71. Hohenberg P, Kohn W (1964) Inhomogeneous electron gas. Phys Rev 136:B864–B871
72. Kohn W, Sham LJ (1965) Self–consistent equations including exchange and correlation effects. Phys Rev 140:A1133–A1138
73. Clarkson C, Ellis G, Larena J, Umeh O (2011) Does the growth of structure affect our dynamical models of the Universe? The averaging, backreaction, and fitting problems in cosmology. Rep Prog Phys 74:112901
74. Wilson KG (1979) Problems in physics with many scales of length. Sci Am 241:158–179
75. Nicolis G, Prigogine I (1977) Self–organization in nonequilibrium systems: From dissipative structures to order through fluctuations. Wiley, New York
76. Haken H (1983) Synergetics, an introduction: nonequilibrium phase transitions and self–organization in physics, chemistry, and biology. Springer, New York
77. Xu J, Han M, Ren Y, Li J (2013) Exploring the mechanisms of protein folding. arXiv:1307.4147v2
78. Turing AM (1952) The chemical basis of morphogenesis. Phil Trans R Soc London B 237:37–72
79. Huang W (2012) Personal communication. Institute of Process Engineering, Chinese Academy of Sciences
80. Edwards PP, Johnston RL, Rao CNR (2000) On the size–induced metal–insulator transition in clusters and small particles. In: Braunstein P, Oro LA, Raithby PR (eds) Metal clusters in chemistry, chap 4.8, Wiley, New York
81. Fujita T (1959) Precipitation and cold air production in mesoscale thunderstorm systems. J Meteorol 16:454–466
82. Orlanski I (1975) A rational subdivision of scales for atmospheric processes. Bull Am Meteorol Soc 56:527–530
83. Zhang Y, Bocquet M, Mallet V, Seigneur C, Baklanov A (2012) Real–time air quality forecasting, part I: history, techniques, and current status. Atmos Environ 60:632–655
84. http://en.wikipedia.org/wiki/Mesoscale_meteorology
85. Fan J, Boettcher SW, Tsung C–K, Shi Q, Schierhorn M, Stucky GD (2008) Field–directed and confined molecular assembly of mesostructured materials: basic principles and new opportunities. Chem Mater 20:909–921
86. Saksena RS, Coveney PV (2008) Self–assembly of ternary cubic, hexagonal, and lamellar mesophases using the lattice–Boltzmann kinetic method. J Phys Chem B 112:2950–2957

87. Lyakhova KS, Horvat A, Zvelindovsky AV, Sevink GJA (2006) Dynamics of terrace formation in a nanostructured thin block copolymer film. Langmuir 22:5848–5855

88. Ramanathan M, Darling SB (2011) Mesoscale morphologies in polymer thin films. Prog Polym Sci 36:793–812

89. Schmalian J, Wolynes PG (2005) Electronic mayonnaise: uniting the sciences of "hard" and "soft" matter. MRS Bull 30:433–436

90. Walsh C (1998) Chemical biology: information, mesoscale science and the engineering ethos. Chem Biol 5(8):R177–R179

91. Boettcher SW, Fan J, Tsung C–K, Shi Q, Stucky GD (2007) Harnessing the sol–gel process for the assembly of non–silicate mesostructured oxide materials. Acc Chem Res 40(9):784–792

92. Xie R, Feng Z, Li S, Xu B (2011) EDTA–assisted self–assembly of fluoride–substituted hydroxyapatite coating on enamel substrate. Cryst Growth Des 11:5206–5214

93. http://en.wikipedia.org/wiki/Meso_compound

94. Lehner E (2007) Describing students of the African Diaspora: understanding micro and meso level science learning as gateways to standards based discourse. Cult Stud Sci Edu 2:441–473

95. Förster S, Antonietti M (1998) Amphiphilic block copolymers in structure–controlled nanomaterial hybrids. Adv Mater 10:195–217

96. Brännström Å, Sumpter DJT (2005) The role of competition and clustering in population dynamics. Proc R Soc B 272:2065–2072

97. Whitesides GM, Grzybowski B (2002) Self–assembly at all scales. Science 295:2418–2421

98. Ball P (1999) The self–made tapestry: pattern formation in nature. Oxford Univ Press, Oxford

99. Anderson PW (1972) More is different: broken symmetry and the nature of the hierarchical structure of science. Science 177(4047):393–396

100. http://www.nsfc.gov.cn/Portal0/InfoModule_584/50112.htm

101. Service RF (2012) The next big(ger) thing. Science 335:1167

# Index

J. Li and W. Huang, *Towards Mesoscience*, SpringerBriefs in Applied
Sciences and Technology, DOI: 10.1007/978-3-642-41790-0,
© The Author(s) 2014